Lightning-Fast Mobile App Development with Galio

Build stylish cross-platform mobile apps with Galio and React Native

Alin Gheorghe

 GalioPress

GALIOPRESS IS AN IMPRINT OF PACKT PUBLISHING

Lightning-Fast Mobile App Development with Galio

Associate Group Product Manager: Pavan Ramchandani
Publishing Product Manager: Aaron Tanna
Senior Editor: Sofi Rogers
Content Development Editor: Rakhi Patel
Technical Editor: Simran Udasi
Copy Editor: Safis Editing
Project Coordinator: Manthan Patel
Proofreader: Safis Editing
Indexer: Manju Arsan
Production Designer: Sinhayna Bais

First published: September 2021

Production reference: 1300921

Published by Packt Publishing Ltd.
Livery Place
35 Livery Street
Birmingham
B3 2PB, UK.

ISBN 978-1-80107-316-5

www.packt.com

To my mother, Elena Daniela, and my father, Mircea Bogdan, who have supported me ever since I first displayed an interest in technology and art. I know you always tried your best to give me the best childhood I could possibly have, and because of that, I truly appreciate you.

I'm still thinking about you, whose words and love motivated me to surpass my limits.

– Alin Gheorghe

Contributors

About the author

Alin Gheorghe is a web developer based in Bucharest, Romania. After he created his first iOS game at the age of 15, he fell in love with the world of mobile applications. He has worked with a wide range of different technologies while also developing his love of music and photography. He truly believes coding is just another form of art that requires the developer to think outside the box.

> *I want to thank the people who have been close to me and supported me. I wouldn't have been able to get to this point without your love and support.*

About the reviewer

Rommel Rico is a software engineer from San Diego, California. A naturally curious individual, he enjoys learning new technologies and discovering the best way to solve a given problem across a multitude of platforms and languages. He co-founded Fiber Club, a re-imagined fiber platform, where he is in charge of all the technical aspects of the project. When he is not busy tinkering with code, you might see him on his bicycle riding around town, attending a soccer game, or practicing the Korean language. Rommel identifies as Mexican-American and uses he/him pronouns. You can follow him on Twitter @rommelrico.

Table of Contents

3
The Correct Mindset

4
Your First
Cross-Platform App

5
Why Galio?

6
The Basics of Mobile UI Building

7

Exploring the State of Our App

8

Creating Your Own Custom Components

9

Debugging and Reaching Out
for Help

10
Building an Onboarding Screen

11
Let's Build – Stopwatch App

12
Where To Go from Here?

Other Books You May Enjoy

Index

Preface

This book is the definitive guide to Galio mobile app development and shows you how to set up React Native projects for your own ideas. With the help of step-by-step explanations of essential concepts and practical examples, this book helps you to understand the basics of React Native and how Galio works.

Who this book is for

This book is for developers who are looking to learn new skills or build personal mobile apps. Anyone trying to change their job as well as beginners and intermediate web developers will also find this book useful. A basic understanding of CSS, HTML, and JavaScript is needed to get the most out of this book.

What this book covers

In *Chapter 1*, *Introduction to React Native and Galio*, you will learn about the power of React Native. There will be an easy introduction to what React Native is, and you will discover where Galio comes in and how it saves you time and stress.

In *Chapter 2*, *Basics of React Native*, you will learn about the basic concepts of React Native, such as JSX and the basic components this framework has to offer. You will also find out more about the correct directory structure for your apps and how you can make the most out of it.

Chapter 3, *The Correct Mindset*, deals with the way any user should be looking at working with React. This will help you pick up some good habits for developing mobile apps and software. It also serves as a transition between the basics and actually creating your first cross-platform mobile app.

In *Chapter 4*, *Your First Cross-Platform App*, you will learn how to create your first cross-platform app through a hands-on example. This chapter is meant to serve as an introduction to packaging, how to use npm, and why Galio is needed.

In *Chapter 5, Why Galio?*, we'll go over the things Galio does best, how it can help you, and how reaching out and helping the community can benefit you. This will inspire you to be a productive member of the open source community and learn more about React Native.

Chapter 6, The Basics of Mobile UI Building, helps you understand the basics of constructing a basic but beautiful UI for an app. You are probably sick of looking at ugly apps, and given the chance, you would want to create something beautiful. This chapter is all about how you can do that.

In *Chapter 7, Exploring the State of Our App*, you will see how the components behave side by side and understand how, why, and where to use Galio components. Doing this will help you develop your own way of critical thinking.

Chapter 8, Creating Your Own Custom Components, will teach you how to build your own components based on Galio. You will discover how to combine those already existing beautiful components into the ones that you need in your app.

Chapter 9, Debugging and Reaching Out for Help, will teach you how to debug your own app and reach out for help when you need it.

In *Chapter 10, Building an Onboarding Screen*, you will begin creating React Native apps; I've chosen the onboarding screen because it is usually the first screen you see when you open up an app.

In *Chapter 11, Let's Build – Stopwatch App*, you will learn how to combine your first screen and use React Navigation in order to connect it with the Stopwatch screen. This screen will be a bit more difficult because it has a real use case, but that will make things more rewarding.

Chapter 12, Where to Go from Here?, is where you will learn some more about React Native, Galio, and how to transform yourself in order to be a great and successful mobile developer.

To get the most out of this book

I'm assuming you will have beginner-level knowledge of HTML, CSS, and JavaScript. Having some experience with React would definitely be an advantage but it is not necessary. You will need a Windows/Mac computer with the latest software installed.

Software/hardware covered in the book	Operating system requirements
React Native 0.65	Windows, macOS, or Linux
React Navigation v5	
Expo 41 and 42	
Galio 0.8	
Node.js 14.17.6 LTS	
Android Studio (>=4.1)	
Xcode (>=9.4)	macOS
Homebrew	macOS
Chocolatey	Windows

If you are using the digital version of this book, we advise you to type the code yourself or access the code from the book's GitHub repository (a link is available in the next section). Doing so will help you avoid any potential errors related to the copying and pasting of code.

After reading the book, I'd like you to try to re-do all the challenges in the book all by yourself without looking at any code, all the while adding a personal touch of your own to each and every exercise.

Download the example code files

You can download the example code files for this book from GitHub at `https://github.com/PacktPublishing/Lightning-Fast-Mobile-App-Development-with-Galio`. If there's an update to the code, it will be updated in the GitHub repository.

We also have other code bundles from our rich catalog of books and videos available at `https://github.com/PacktPublishing/`. Check them out!

Download the color images

We also provide a PDF file that has color images of the screenshots and diagrams used in this book. You can download it here: `https://static.packtcdn.com/downloads/9781801073165_ColorImages.pdf`.

Conventions used

There are a number of text conventions used throughout this book.

`Code in text`: Indicates code words in text, database table names, folder names, filenames, file extensions, pathnames, dummy URLs, user input, and Twitter handles. Here is an example: "Now, for the second row, let's go inside our `styles.row2` object and add padding."

A block of code is set as follows:

```
const styles = theme => StyleSheet.create({
  container: {
    flex: 1,
    backgroundColor: theme.COLORS.FACEBOOK
  }
});
```

Any command-line input or output is written as follows:

```
npm i galio-framework
```

Bold: Indicates a new term, an important word, or words that you see onscreen. For instance, words in menus or dialog boxes appear in **bold**. Here is an example: "After writing down your username and password, you should get the following response: **Success. You are now logged in as YOUR-USERNAME.**"

> Tips or important notes
> Appear like this.

Get in touch

Feedback from our readers is always welcome.

General feedback: If you have questions about any aspect of this book, email us at customercare@packtpub.com and mention the book title in the subject of your message.

Errata: Although we have taken every care to ensure the accuracy of our content, mistakes do happen. If you have found a mistake in this book, we would be grateful if you would report this to us. Please visit www.packtpub.com/support/errata and fill in the form.

Piracy: If you come across any illegal copies of our works in any form on the internet, we would be grateful if you would provide us with the location address or website name. Please contact us at copyright@packt.com with a link to the material.

If you are interested in becoming an author: If there is a topic that you have expertise in and you are interested in either writing or contributing to a book, please visit authors.packtpub.com.

Share Your Thoughts

Once you've read *Lightning-Fast Mobile App Development with Galio*, we'd love to hear your thoughts! Scan the QR code below to go straight to the Amazon review page for this book and share your feedback.

https://packt.link/r/1801073163

Your review is important to us and the tech community and will help us make sure we're delivering excellent quality content.

1

Introduction to React Native and Galio

Let's start by understanding what this book is about and how it can help you learn how to use React Native and Galio. By reading this book, you'll understand how to install React Native and all the necessary tools for using it with both macOS and Windows. You will then understand how to create an Expo project and why we're using Expo, the difference between the template workflows and how they come in handy, and also how to start your new project on both a physical device and a simulator. Things should be really easy to follow, so you might find the experience rewarding.

Understanding the world of cross-platform mobile programming isn't an easy task but it sure is a doable one. You made the first step by purchasing this book – the second one is currently in progress since you're reading this book to discover how React Native works and how Galio is meant to help you build apps faster. The main purpose of this book is for you to get accustomed to how React Native works, how to use it for your projects, and how Galio comes in handy and could potentially save a lot of your time.

I can understand this might not be an easy task at first but I strongly recommend going over each section as many times as needed. If there is something that might not be entirely easy to understand at first, you can always ask questions in places such as Stack Overflow or different subreddits. We'll look at places to ask for help in depth later in this book.

At first, most programmers, myself included, thought cross-platform mobile programming frameworks might be much slower than the native ones. This was just a thought as we're going to see that React Native is a really good way of creating mobile apps as they're not at all that slow compared to the native ones.

As you'll understand soon enough, this book is strongly connected to Galio, which I believe to be one of the best-looking UI libraries out there. Galio is going to help us build a React Native app faster and with more style than we would've been able to do on our own.

You'll also learn lots of ways to develop your own UIs and how to start thinking outside of the box while developing your applications. This is important as it could make the difference between a successful app and an unsuccessful one.

Learning the basic rules of design and programming is only the first step in the process of being a complete frontend developer. Learning how to break those rules will develop your skills even further.

Sometimes, there are going to be tips showing up in places where they're most needed, and following them will benefit anyone trying to get into a programmer mindset.

At the end of this book, you'll find exercises and lots of tips on how to develop a more complex UI for your mobile applications. All these have a great purpose in mind and that is to develop a programming style while having a good base.

I strongly believe that by the end of this book, anyone should be able to create at least a basic cross-platform mobile app that'll serve as a good MVP for a personal project. Learning and experiencing all that is written in this book should play an important role not only for you as you learn to use React Native and Galio but also for you as a programmer.

This chapter will cover the following topics:

- Why React Native?
- Galio – the best UI alternative
- Configuring your React Native environment
- Creating your first React Native project

Why React Native?

So, you may be wondering, "Why React Native?". There are so many cross-platform frameworks available out there such as Flutter, Xamarin, and Cordava, to name a few, so we need to understand why React Native is a great choice for mobile application development.

You need to understand that there is no right choice. This is only based on the current context of the market and personal appreciation.

Programming frameworks are like a painter's brush. A painter has multiple brushes, each of them with a different purpose. You need to understand as much as you can about the tool that you're using because the better the painter knows the brush, the better they can paint and bring their vision to life.

You need to learn how to use React Native to develop cross-platform apps quickly and easily. So, let's go a bit more in depth into why React Native is such a great choice for app development.

You only learn it once

First of all, React Native is based on React, which means you only learn it once and you can develop everywhere. This is a really good way of expanding your skills. By learning React Native, you'll be ready to write React code for your website.

This should make it easy for you to understand why this is such a great choice. Imagine that you have already created an app. Your application is cool – people are starting to download it from the App Store or Google Play – but what would help even more is a landing page. Because you've already learned React Native, using your skills with React is going to be a piece of cake.

Bigger talent pool

Back in the old days of programming, when you had an app idea and you wanted to develop it, you had to search for a backend developer with some C# or Java skills, an iOS developer with Objective-C skills, an Android developer that had to know Java, and maybe even some web frontend developers for your application's website.

This requires a lot of effort and a pretty big budget. At the end of the project, your idea might not work in today's market, and you will have lost a lot of time and money.

Now, all those specific jobs could be handled by JavaScript engineers – we have multiple alternatives to the native frameworks that work just as good but they're written in JavaScript, which is one of the most used languages right now. JavaScript developers are even more available on the market and transferring from one framework to another is easier than ever. By hiring a JavaScript developer, the budget is cut in half, the app is developed even faster, and they can help each other even though they have different jobs.

JavaScript developers can easily change teams. A backend developer can help a frontend developer or even the mobile app team. They can go and help out wherever you need more manpower to push the development even faster. This is especially valuable when one of your developers is missing due to resignation or illness.

Having a bigger pool of people to choose from is a big plus for any app development.

React's popularity

You'd think that React's popularity has nothing to do with React Native but actually, React and React Native are close in terms of writing code and methodology. My recommendation is to always look at Google Trends because it can help us understand whether a framework is popular or not:

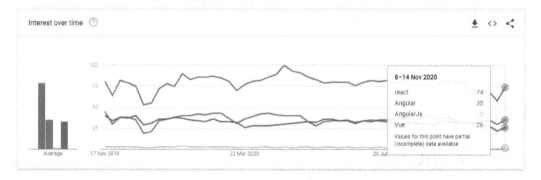

Figure 1.1 – Google Trends showing the popularity of React at this moment

React makes it easy for developers to build a great web UI, but the components-based approach also makes the app a lot easier to maintain. React Native brings all those advantages to the mobile app development universe.

So, what does this show us? There's a pretty big community of people searching for React, and React Native has one of the biggest and most active communities. For almost every little issue that you might encounter, there's someone that has already written an article or opened up an issue for it on GitHub. The community is really big on GitHub as well, which will be helpful as you can get in contact with more developers, ask for help regarding any of the libraries that you might be using inside your app, and get more help for any open source ideas that you might have, which could potentially help a lot of developers.

I recommend that everyone contributes to open source projects because this will help develop your skills and expand your way of thinking like a programmer. The community is so helpful and friendly that you might find it hard to switch to any other framework as this seems like the best choice for most of your needs.

Performance

React Native gets close to a native app from a performance standpoint but you have to use it in the right way. Technically, you have a JavaScript thread running, which is pretty slow compared to native code such as Kotlin for Android or Swift for iOS.

React Native shines because it creates a bridge between the JavaScript thread and the native one. It was designed to move the most expensive and powerful tasks such as rendering to the native side. This is used asynchronously so that the JavaScript thread will not have to wait on the native computations.

Let's say the user presses a button – React Native will translate this into an event that JavaScript can handle. After that, by sending messages between a native platform such as iOS or Android and a JavaScript code base, the React Native bridge translates native events into ones that React components can understand and respond to.

There are certain challenges here, such as default components – which are built-in elements provided by React Native – not looking or responding the same on both platforms since there are so many platform-specific events. There's nothing to worry about though because this bridging architecture allows us to use all the existing native views from platforms, SDKs, and JavaScript libraries.

The language

JavaScript was created as a client-side language. It was built to make websites interactive. If you think of a basic website layout, you have your HTML, which describes the basic content and the website's structure, and then you have your CSS, which styles the HTML and makes it pretty. This is a static website that doesn't do much, so we needed a programming language that could add functionality to our website and bring it to life. This is where JavaScript came into the game.

Time passed and people realized they could do a lot more with JavaScript. The most popular use of JavaScript is client side but ever since Node.js appeared in the programming scene, the language has evolved so much that this is not the case anymore. JavaScript is now an all-purpose programming language, meaning you can use it to build almost anything. You can even have typed JavaScript with TypeScript or Flow. The support inside code editors got a lot better as well.

Having said all this, React Native uses JavaScript as its main programming language. *As we learn more, we'll see that React Native can also use native code to run faster and do better computations.*

Stack Overflow (one of the biggest communities of programmers) does a survey each year where they try to understand more about developers and the people using their platform. You can ask any developer about their platform and almost anyone could tell you they've browsed it at least once. Their 2020 study showed that almost 70% of their user base was formed of professional developers using JavaScript.

Being such a versatile language, learning it for React Native or another framework can only help you expand your territory as a programmer. The fact that React Native uses it is a big plus as it shows how easy this allows you to move between technologies.

You can learn more about Stack Overflow statistics regarding the 2020 study by going to `https://insights.stackoverflow.com/survey/2020`.

Drawing conclusions

After reading all this about React Native, we need to understand that even though React Native is not exactly as fast as a native app, it can be almost as fast as one. And if we throw in the fact that the language has so many opportunities for us developers and also that the community is so strong and friendly, we might see React Native as one of the best frameworks for cross-platform mobile app development.

To pick the library that suits your needs, you need to take what's the most important to you into account. I hope you know a little bit more about React Native and that you're confident that this framework is a good choice for **you**.

Next, we'll understand more about what a UI library is and how Galio comes in like a great sidekick to help us out when we're writing our code.

Galio – the best UI alternative

So, you've learned a thing or two about how React Native works and now you're wondering how Galio can help you. Well, first of all, what exactly is Galio?

To put it simply, Galio is a React Native UI library, so it's a collection of resources meant to help developers write code faster and easier. The thing is... React Native doesn't have that many components. We'll come back to what exactly a component is later in this book, but for now, just think of them as puzzle pieces.

React Native has a certain amount of puzzle pieces, each of which is as simple as possible. Galio comes in as a wrapper around those puzzle pieces and adds a bit of color and functionality. Sometimes, you can even find different pieces that are built by combining more basic pieces into a really big one for specific reasons.

Now, let's go over the reasons why Galio might be the best UI library for you in your cross-platform mobile development journey.

Time-effective

All right, too many metaphors. The thing is, React Native only has basic-looking components, which makes developers build their own. This is time-consuming as you would always have to build new components for your new app.

This is where Galio comes in handy! Packed with a lot of already beautiful components, it eases the pain of creating your own all the time.

Also, all the components are easier to customize and still fit the whole design layout, without putting too much pressure on the developer to think about how to do it and where to start. The process of customizing a component from Galio is straightforward and usually, it revolves around using props, which make the whole thing a lot more readable.

I know that words such as "component" and "props" are completely or maybe just somewhat alien to you right now, but all that matters is that they save you a lot of time. We'll learn more about these keywords shortly, but we need to understand a bit of the overall image of what all these technologies mean in the grand scheme of things.

Building an app with Galio is usually more about the way you choose to create your layout than it is about actually programming the UI. It is using a direct way of creating a mobile screen by placing each component under the previous one. This allows us to be more efficient and waste as little time as possible as we code and think about what the final screen should look like.

The following diagram shows the basic program structure you can create using the pieces of the puzzle we were discussing:

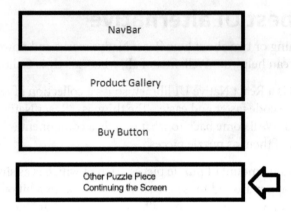

Figure 1.2 – Representation of how adding more components helps build a mobile screen

This is a great way of thinking because it makes you understand some of the best principles of atomic design. It also creates a more organized code base that you can expand into a more complex and fuller app.

It's really beautiful

The fact that Galio comes already packed with a design system in mind means that all the components will follow the same design principles, so there's never going to be any discrepancies between the components.

A consistent design is what makes an app complete. A consistent design will help the user understand your app flow better, all the symbols that you wish to incorporate in your app, and how to use it. It's all about being consistent with your buttons, texts, and design.

You might not necessarily like the colors at first, which is OK because you can always change them easily with the help of <GalioProvider>. We will look at this later in this book.

Right now, we've learned about why React Native is such a great choice and why Galio is a great UI library for us to start building apps. The next step is to understand how to configure a good environment so that we can start building cross-platform mobile apps.

Configuring your React Native environment

There are two things that we need to know about: **Expo** and **the React Native CLI**.

Both of them are easy to install and we'll go over both of them to make sure we cover all possible ground. I recommend not skipping over this part as it will help you make a good choice when you start developing your mobile project.

Things are going to be much easier to install on macOS compared to Windows as macOS is a UNIX-based system, so the terminal is a lot more powerful. But worry not – we will solve this problem for Windows as well.

We have some requirements we must consider before we move on. These are going to help us create a good environment for both Expo and the React Native CLI, and also for JavaScript programming in general.

We'll need the following technologies installed for either system:

- Homebrew – macOS only
- Chocolatey – Windows only
- Node.js
- Text Editor
- Android Studio
- Xcode – macOS only

We'll start by installing Node.js, which is one of the most important technologies we need for JavaScript to run outside the browser. Node.js is built on Chrome's V8 JavaScript engine, which means you would be able to run any JavaScript code that runs on the latest Chrome version (the web browser).

The recommended way of installing Node.js depends on your operating system. For macOS users, the best way would be by using Homebrew, while for Windows users, you would use Chocolatey. Homebrew and Chocolatey are package managers that enable you to install different packages such as Node.js easier and faster, all via the command line or terminal. You can also install it via the official website at `https://nodejs.org`, but we're going to use Homebrew or Chocolatey in this book.

Homebrew

For macOS, we have Homebrew, which is easy to install. You can find more information on it on their official website at `https://brew.sh`.

To install it, you should open **Terminal** and write the following command:

```
/bin/bash -c "$(curl -fsSL https://raw.githubusercontent.com/
Homebrew/install/master/install.sh)"
```

After writing the command, press *Enter*. More information regarding everything that is going to be installed will appear; just press *Enter* again and you should be ready to go.

Chocolatey

For Windows, we have Chocolatey, which is a bit more complicated than Homebrew to install but by following the steps here, you should be all set. You can find out more about Chocolatey by going to their official website at `https://chocolatey.org`.

First of all, we need to use PowerShell with administrative rights. All you have to do to access it is to press the *Windows logo + X* on your keyboard. A new menu will appear in the bottom-left corner of your screen. Here, select **Windows Powershell (Admin)**. A new window will open.

First, you need to verify that `Get-ExecutionPolicy` is not Restricted, so write the following in PowerShell:

```
Get-ExecutionPolicy
```

If it returns `Restricted`, then you need to run the following command:

```
Set-ExecutionPolicyAllSigned
```

Now, you're all set to run the following command:

```
Set-ExecutionPolicy Bypass -Scope Process -Force; [System.
Net.ServicePointManager]::SecurityProtocol = [System.Net.
ServicePointManager]::SecurityProtocol -bor 3072; iex ((New-
Object System.Net.WebClient).DownloadString('https://
chocolatey.org/install.ps1'))
```

Now, wait for a few minutes for everything to be installed. If you didn't encounter any errors during installation, just type `choco` to return your Chocolatey version. If it does return it, then you're all set and ready to go.

All we need to do right now is install Node.js so that we can learn about Expo and the React Native CLI. Having installed Homebrew or Chocolatey makes this easy as all you have to do is write the following commands and Node.js will start installing:

- Use the following command on macOS:

  ```
  brew install node
  ```

- Use the following command on Windows:

  ```
  choco install -y nodejs
  ```

Congratulations! We're now ready to move forward! With that, we have installed Node.js. Before setting up our environment, let's discuss text editors for a second – I promise you it won't take long.

I bet you're thinking *Wait, did he say we can write code in a Word document?*. Well, not really. Microsoft Word is not a plain text editor, but you can use something such as Notepad to write code. Just because we can use Notepad doesn't mean we will use it; it doesn't look very professional, does it?

The type of text editor we're going to use is going to have cool features such as a color scheme for our code syntax, as well as different add-ons that will help us write code faster and prettier.

There are many different free text editors out there, including Sublime, Atom, Visual Studio Code, Notepad++, and Brackets. They're all equally good and I recommend that you download at least two or three of them and check them out. My personal preference is Visual Studio Code, and I am going to use it throughout this book. You don't need to use the same text editor if you don't like the way it looks because you can follow this book by using any of the aforementioned editors.

You can download Visual Studio Code (or just VSCode) from https://code.visualstudio.com/.

Now that we have got some necessities out of the way, it is time to move on and learn about Expo and the React Native CLI. Both of them can be used to get to the same outcome – they're just different ways of creating a React Native app and we'll try to understand them as much as we can. Knowing everything about them is going to help us choose the right one for us and our app.

The React Native CLI

The React Native CLI is the official and first method of creating a React Native project. It is usually harder to configure and it takes a lot more time than Expo, but it is worth it. After all, you need a simulator to test your app on different phones. I recommend not skipping this section.

macOS

A good thing about having macOS is the fact that you can simulate an iPhone and see what your project looks like on different Apple technologies. This is something that you can't do on Windows but Android works on both, so macOS has an advantage in being able to simulate all types of platforms.

We should get going and install all of the necessary dependencies; open Terminal and write the following:

```
brew install watchman
```

Watchman is a tool developed by Facebook to watch the changes inside the filesystem. It also offers better performance.

Now, you need to install Xcode. Go ahead and open the Mac App Store, search for Xcode, and click on **Install**. This will also install the iOS Simulator and the rest of the tools we need to build our iOS apps. Your Xcode version needs to be at least 9.4 for React Native to work with it.

Now, we need the Xcode Command Line Tools package. Once Xcode has finished downloading and installing, open it up and go to **Preferences** (it's under the Xcode menu in the navbar; alternatively, just press *Cmd + ,*). A new window should open. Go to **Locations** and install Command Line Tools by selecting the most recent version from the dropdown:

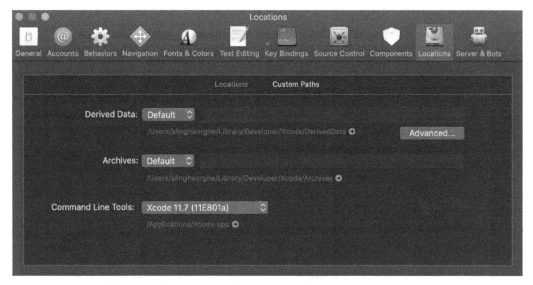

Figure 1.3 – Preferences window in Xcode

Now, go to the **Components** tab and install the latest simulators that you wish to use.

> **Important Note**
> There is a simulator for each iOS version supported by Apple. You should try
> to install the last two major versions since your app users might always be on a
> previous iOS version.

All you need to do now is install CocoaPods by writing the following in Terminal:

```
sudo gem install cocoapods
```

This is a Ruby gem that manages dependencies for your Xcode projects.

With that, you're all set to create your first project on macOS! We'll do this in just a bit!

Windows

As we know, we can't install any simulators for iOS on Windows, so we might as well just
install the Android one.

We have already installed Node.js, so all that's left to do is install JDK by going to our
admin PowerShell (we explained how to open it earlier when we installed Node.js and
Chocolatey). Once you've opened it, write the following:

```
choco install -y openjdk8
```

If you already have JDK installed, make sure it's at least v8.

Now, it's time to install our Android development environment, which can be a bit tedious. However, it's worth it as we're going to be able to run our React Native apps right on our virtual Android simulator.

Go ahead and download Android Studio by going to https://developer.android. com/studio. Once it has finished installing, start Android Studio. Once you've opened it, choose the theme that you like and all the preferences that are good for your computer. At one point, the **SDK Components Setup** page will appear. Make sure you have selected the **Android SDK**, **Android SDK Platform**, and **Android Virtual Device** checkboxes.

Once the installation has been completed, it is time to move on. Android Studio is always installing the latest Android SDK by default. However, building a React Native app with native Android code requires the **Android 10 (Q) SDK**. To install it, open Android Studio, click on **Configure** at the bottom-right corner of the window, and select **SDK Manager**:

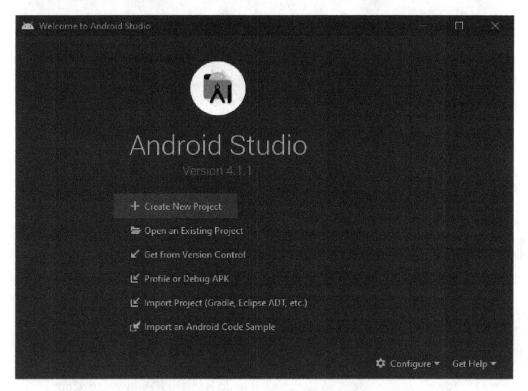

Figure 1.4 – Android Studio and the positioning of the buttons

Now, select the **SDK Platforms** tab and check the box in the bottom-right corner that states **Show package details**. Look for and expand Android 10 (Q) and make sure the following are checked:

- **Android SDK Platform 29**

- **Intel x86 Atom_64 System Image or Google APIs Intel x86 Atom System Image**

The next thing you should select is the **SDK Tools** tab and check the box next to **Show Package details**. Look for **Android SDK Build-Tools** and select **29.0.2**.

Click **Apply** and download all the necessary files.

Now, it's time to configure the ANDROID_HOME environment variable so that we can use native code. Open **Control Panel**, click on **User Accounts**, and then on **User Accounts** once more. On the left-hand side, you'll find **Change my environment variables**; click on it. Now, click on **New...** and write the following:

- **Variable name**: ANDROID_HOME.

- **Variable value**: C:\Users\{name}\AppData\Local\Android\Sdk, where {name} is the username of your PC:

Figure 1.5 – Windows showing my user variable

To check whether the environment variable has been loaded, go to your PowerShell environment and write the following:

```
Get-ChildItem -Path Env:\
```

You should see a list, and ANDROID_HOME should be part of it.

After that, we need to add `platform-tools` to **Path**. We can do that by going to **Control Panel**, clicking **User Accounts**, and then clicking on **User Accounts** once more. Click on **Change my environment variables** and look for **Path**. Select **Path** and click **Edit**. A new window will appear where we can click **New**. Now, use the same variable value that we did previously but this time, go into the `Sdk` folder – more precisely, the `platform-tools` folder:

`C:\Users\{name}\AppData\Local\Android\Sdk\platform-tools`

We're now ready to start developing React Native apps on Windows!

Expo

Expo is the easiest way to start a React Native project for a beginner. It comes packed with a big set of tools built for React Native, but we're not interested in that right now. We're only interested in the fact that Expo can get you up and running in a few minutes and it doesn't require you to have a simulator installed, so you can play with your app in a matter of minutes.

They also came up with something called Snack (`https://snack.expo.io/`), which is helpful if you wish to try out different code ideas directly in your browser! This is cool as you don't even need to start a project if just want to sketch something quickly.

Let's install it and see whether this is as easy as I make it out to be. Open a terminal or command line and write the following:

```
npm install -g expo-cli
```

Expo is now ready to go! Easy, right?

Ready to go even further?

Now that we've installed all the necessary technologies, we're ready to create our own React Native project and create some awesome apps!

But first, let's understand the difference between the React Native CLI and Expo. Previously, I told you not to skip the React Native CLI part, even though it's a lot bigger than the Expo one. That's because we need to install Xcode or Android Studio to have some control over our app directly from our PC.

We haven't created a project with either the React Native CLI or Expo yet because they're both created differently. However, we've installed the requirements for both of them. Creating a project with the React Native CLI leaves the developer with the job of fully creating an app from 0. You have complete control over the app and there's nothing but your imagination that can limit what you can do. You can even use native code – Kotlin/Java for Android or Swift/Objective-C for iOS – and create your own completely native components. This is all very advanced, though, and we don't need it.

Expo comes packed with a lot of tools for people who want to create a fast and powerful app, without the struggle of having to think about all the little details regarding how the app works and connects with each specific platform.

For this reason, we'll use Expo to create the projects in this book.

Creating your first React Native project

We're ready to go! Let's open up a terminal and head right into it!

Once the terminal is open, just move to your `Desktop` folder or any folder you wish to create your project in. You can move between folders by using the `cd` command. So, just by writing `cd Desktop`, we've arrived at the **Desktop** directory and we're ready to create our Expo project.

We can create a new React Native project with Expo by writing the following:

```
expo init MyFirstProject
```

Upon pressing *Enter*, Expo tells us we can choose between a multitude of templates. The biggest two categories are **Managed workflow** and **Bare workflow**:

```
C:\Users\yovng ra\Desktop>expo init MyFirstProject
? Choose a template: (Use arrow keys)
 ----- Managed workflow -----
> blank                 a minimal app as clean as an empty canvas
  blank (TypeScript)    same as blank but with TypeScript configuration
  tabs (TypeScript)     several example screens and tabs using react-navigation and TypeScript
 ----- Bare workflow -----
  minimal               bare and minimal, just the essentials to get you started
  minimal (TypeScript)  same as minimal but with TypeScript configuration
```

Figure 1.6 – A representation of what you will see after initializing a project

We'll explain what both are in a few seconds. For now, pick **blank** under **Managed workflow**. Wait a few seconds and you'll see a message in your terminal stating the following:

```
Your project is ready!
```

Now, we're ready to start it. Write the following command in the terminal; this will move us into our project's folder:

```
cd MyFirstProject
```

Now that we're here, let's understand how each type of template works so that we can start playing with our first React Native app.

Managed workflow

The managed workflow tries to take care of almost all the complex actions you would have to do. This is usually for complete beginners who don't want to complicate things with Xcode or Android Studio, which is exactly why we've started by creating a project with this type of workflow. You can change all the app's information, such as its icon or splash screen, through app.json and have direct, easy access to different services such as push notifications.

This workflow has some limitations, though. Let's say you want to use a specific device capability that the Expo SDK doesn't offer through their API. This means that you would need to run eject and expo-cli will do all the work and transfer you to the **bare workflow**.

Bare workflow

The bare workflow presents the developer with complete control over the app. However, this comes with all the complexity of having to take care of every little detail regarding your app. Now, the ability to easily configure app.json has disappeared and you're left with only Expo SDK installed and no pre-configuration done.

This allows you to use native code and manage your dependencies however you want. You might be wondering, "*Well... isn't this the same thing as using the React Native CLI?*". Well, not really, because you get instant access to the Expo SDK and Expo framework, which within itself represents a big plus for a developer as it still eases your development process.

Opening our project files

Now that we've understood what each template does and why we chose the managed workflow, let's see what the code looks like.

Remember when we discussed text editors? Go ahead and open the text editor you chose. I'll use VSCode as I prefer the design compared to the others.

Click on **File | Open Folder** and search for the project's folder. Mine's in the Desktop folder. Opening it lets us see all the files and folders inside our project.

I'm pretty sure you're confused about the purpose of each of these files. We'll look at this shortly, but for now, take a look around for a few minutes and see whether you can pick up anything by yourself.

> Tip
> Investigating any piece of code that you find and create is the best way to ensure you're learning the stuff that you're reading or coding. The best programmers are always the ones that use their deductive skills.

Preparing our physical device for preview

It's time to prepare our phone so that it can preview our app because it's always cool to show off to our friends with our new skills.

So, let's go to the App Store or Google Play and search for Expo. Install it and you'll be ready to go. This app allows us to test our apps right on our phones, so let's test it out!

Go to your terminal and make your way to the project's folder, if you're not already there. Write the following command:

```
npm start
```

After that, press *Enter*. A new window should open in your default browser. A server has been created and your app can now be previewed on your physical device or even a simulator. On the sidebar, there is a QR code, and on top of that, a link and some buttons. What is this and how can we make use of it?

Figure 1.7 – Preview of all the buttons shown on the screen

Now, you can open your smartphone and either scan the QR code or paste the link above the QR code into a browser. A message will appear, asking you whether you're OK with opening the link with Expo. Press **Yes** and there you have it – your first React Native app on your physical device.

Pretty cool, right?

Let's see what happens if we press **Run on Android device/emulator** or **Run on iOS simulator**. A message will appear in the top-right corner, letting you know Expo is **attempting to open a simulator**. Depending on which operating system you've started your project on and which simulator you have installed, choose the appropriate button.

For the Android simulator, you'll have to open Android Studio first. Then, you must go to the top-right corner, where it says **Configure**, and choose **AVD Manager**. Now, a new window will appear, displaying all the available virtual devices. If you don't see any, go to the bottom-left corner and click **Create virtual device**.

At this point, you will see a list of Android devices. I picked the Pixel 3a but you can choose an older one if your CPU is not very strong. After that, click **Next**; you will be asked to pick a system image. Locate the **x86 images** tab next to **Recommended** and pick the image that doesn't need to be downloaded called **Q**. If all of them need to be downloaded, then you need to go back to the Android Studio installation part and repeat that process. After selecting the image, click **Next** and name the AVD; it can be called anything, so be creative. Click **Finish**; you should see your device in the **AVD Manager** list. On the same row, to the right, you'll find a little green button that looks like the **play** symbol. Click it.

A new Android Simulator will open, so let's head back to the browser tab where we have our Expo server running and click on **Run on Android device/emulator**. If you look inside the terminal, you'll see that some writing has appeared. We just need to wait a second for the Expo client to be downloaded and installed on our simulator. It should say something like this:

```
Downloading the Expo client app [================================
====================================] 100% 0.0s
Installing Expo client on device
Opening exp://192.168.1.111:19000 on Pixel_3a_API_30_x86
```

Now, your Android simulator should show a preview of your first React Native app. How does it feel? You've been through quite a lot but in the end, being able to correctly initialize a project is quite rewarding:

Figure 1.8 – Android Simulator displaying a fresh new React Native project

Before we move on, I think we should learn how to close a project in the terminal/ command line. Go back to the terminal window and click on it so that it's in focus. Now, if you press *Ctrl + C*, the server should stop and you should be able to use that window of the terminal again.

Summary

This chapter started with a brief introduction to React Native and Galio, and then we understood the main focus points as to why those libraries are good for your next personal cross-platform mobile app project. After fully understanding why learning about those libraries will help them become great assets in the future, we started setting up our React Native environment and learning everything there is to learn about testing and utilizing our soon-to-be-created mobile app.

We then created and tested our first React Native app! This has all been quite the journey for a newcomer into this beautiful world of programming and, trust me, it was all worth it. Everything you've learned here will serve as a base for what's coming up next. Isn't this all exciting? In the next chapter, it's time for us to learn the basics of React Native code writing and create even cooler apps.

2
Basics of React Native

We started by learning about why React Native and Galio form the best combination for us to start building our first cross-platform mobile application. After setting up our environment and configuring the necessary files, we created our first React Native project with Expo and we learned about the different ways we can test the app both physically and digitally.

I believe that learning about the whys before the hows helps build a better, more robust knowledge base. Having passed through the whys, it is now time to learn about how React Native works and how to use it to create our apps.

That is why we'll start this chapter by going through the file structure of our React Native project for us to understand how these files and folders are connected. We'll then go through the App.js file and explain how this works for us as the main entry point into our application.

Once we've learned about the file structure, it is time for us to learn about what **JSX** is and how to use it – the skeleton of any React application. We'll be comparing JSX to HTML a lot, so you'll have to know a bit of HTML beforehand. Rest easy if you don't know much about web development – we'll lay down some HTML concepts as well, but learning a bit about it on your own may help you a lot in the long run. *Understanding the concept of JSX* is where we'll deal with the concept of **components**, a concept we barely touched on in the first chapter. This should be completely understood by the end of this chapter.

Once we've understood the main concepts of JSX and how it is connected to React and React Native, we'll do our first component import. We'll learn what npm/yarn is and how to use it for importing and uploading components or libraries on the web. This is exciting as you'll see the importance of having a huge community backing a framework and how you can participate and make new friends.

It is now time to learn about the core components of React Native. We'll understand how and in which contexts they're useful while we discuss different ways of improving them or even changing them completely. The core components are the base components for all the components we'll find online. That means that almost every component inherits from the core ones, which makes them important to learn about and understand.

By the end of this chapter, you'll have learned how to build a component. You'll have also learned how to use it in our app or future apps and how to organize the files so that you'll never get lost searching for your components.

I believe that by the end of this chapter, you will be able to start building really simple apps that can serve as a stepping stone for building bigger, more complex projects. Understanding these concepts doesn't stop at reading this book – it goes further than that, and you'll see me always encouraging you to check out the official documentation of whatever project/framework we're using as the documentation should always be something a programmer should feel comfortable reading. Learning to read the documentation is a skill you'll develop in time by reading and being as interested as possible in the project you're passionate about.

This chapter will cover the following topics:

- Using App.js – the main entry point
- Understanding the concept of JSX
- Importing your first component
- Core components
- Understanding and creating a component

Technical requirements

You can check out this chapter's code by going to GitHub at https://github.com/PacktPublishing/Lightning-Fast-Mobile-App-Development-with-Galio. You'll find a folder called Chapter 02 that contains all the code we've written inside this chapter. To use that project, please follow the instructions in the README.md file.

Using App.js – the main entry point

As we know, React Native is an open source framework used for building iOS and Android applications. It uses React to describe the UI while accessing the platform's capabilities through the methods we have at our disposal.

Understanding our folder structure is important because we're not supposed to touch some of the files– at least at the beginning of our development. Let's take a look at our newly created project's structure.

> **Tip**
>
> Don't forget that you can use any text editor you want to; I'm only using VSCode because I like the way it looks and it has lots of plugins that I use, but that doesn't mean you can't open up the project with whatever text editor you feel comfortable with. Of course, that'll mean you won't be able to use the code . command as that's only used for VSCode.

First, let's open our Terminal and navigate to our folder. Now, if we write code . once we get to the folder, it will open Visual Studio Code.

Once our text editor has opened, we will see the following output in the project directory:

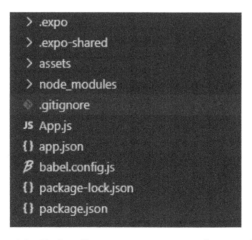

Figure 2.1 – Project directory once you open the text editor

As we can see, there are several folders and files here and they're all meant to help bundle the project once you finish your app. We'll look at each of these folders in the next few sections.

The .expo and .expo-shared folders

We'll start with the folders with a dot in front of them: `.expo` and `.expo-shared`. The dot is there to show a hidden file. That's a file you can't see directly while opening the file browser; you can only see it if you specifically choose to see it. Such files are hidden because you don't need to touch them. They're config files that are created when you first use the `expo start` command.

The assets folder

The next folder is the `assets` folder. Inside, you'll find several `.png` images, which are used by Expo for the splash screen – the screen that appears while the app is loading – and icons for the app to use when it's installed on a device.

The node_modules folder

Now, you'll see a folder called `node_modules`. If you open that folder, you'll be able to see lots and lots of folders. All these folders are packages and **dependencies** that we're using to make this app work. Everything that you're installing or bringing in from over the internet is going to go straight into this folder. This folder is going to get bigger and bigger, depending on how many external packages you'll be using for your app.

Once we get past these folders, we've got some files with some interesting features.

The files within

First, we can see `.gitignore`, which helps us save size when uploading on GitHub. If you open the file, you'll see there's already some text written in it. Everything you see in there will be ignored once you upload your project on GitHub. You'll find that `.expo` is there because those folders are only relevant for the programmer and they're not intended for sharing. You can edit this file by using any filename you don't want to move over to online or you don't intend on changing.

> **Important Note**
> GitHub is a platform that acts like an internet hosting company for open source software programs while also providing the programmer with version control using Git. Developers are using Git to track changes in their projects and coordinating with their teammates.

For now, we'll ignore App.js because we'll explain this file at the end of this section. So, let's go directly to the app.json file. This file acts like a config file for your app – basically, everything that's not code-related will be found there. Let's say, for example, we want to change the image of our splash screen. We don't have any way of doing that besides going into this file and editing the splash image's path. From here, you can change almost everything related to your application, such as the icon or its orientation. You'll see yourself going there quite a lot, configuring your app for the final release version.

We don't care about babel.config.js but I'm pretty sure you'd be curious about that one as well. Babel is a JavaScript compiler almost everyone is using to get access to the latest standards of JavaScript. It's not necessary to edit this file but if you want to learn more about compilers, I recommend searching for more information about Babel.

The last two files are package-lock.json and package.json. The first one always gets created when you're using npm to install dependencies in your project. I've already told you that we'll learn about npm in this chapter, but not right now. Right now, I want you to become familiar with all the files in the project directory. By creating your app via the command line, Expo automatically used npm to bring lots of files you'd use in your project over the internet. Those files are stored in the node_modules folder. You can find out more about all the direct dependencies that you're using in package.json.

Now that we've finally got to the end of all the files, we should start talking about App.js. So, let's open that file and take a look at it.

The App.js file

Upon opening the App.js file, you will see the following:

```
import { StatusBar } from 'expo-status-bar';
import React from 'react';
import { StyleSheet, Text, View } from 'react-native';

export default function App() {
  return (
    <View style={styles.container}>
      <Text>Open up App.js to start working on your app!</Text>
      <StatusBar style="auto" />
    </View>
  );
}

const styles = StyleSheet.create({
  container: {
    flex: 1,
    backgroundColor: '#fff',
    alignItems: 'center',
    justifyContent: 'center',
  },
});
```

Figure 2.2 – Code in the App.js file

You can immediately see the **Open up App.js to start working on your app!** text. I'm pretty sure you remember but in the previous chapter, when we tested our app, this was the text that appeared on the center of the screen. This means that by changing the text, we should also see a change in our app.

We won't do that right now as our focus is understanding the files and code, and then changing it to our liking.

I'm pretty sure you connected the dots after seeing this file and realized that this is the entry point of our app. An entry point is the main file that connects all the files and starts the application. Our main function for using Expo is the App() function. Your entire application will be living inside that function.

The reason you saw the centered text when you opened the app was because the text is inside the App() function. Here, we'll start building our app. For that to happen, we must understand what JSX is and how to use it inside our app. I'm assuming you can already read a bit of JavaScript and you understand notions such as functions and objects; we won't be covering this topic in this book. We'll get to grips with JSX in the next section.

Understanding the concept of JSX

We've finally got here, we're now ready to look over JSX and learn how to use it inside our apps. React is heavily dependent on JSX as it's the main way of building the layout for our apps.

First, we'll take a look at a variable containing some JSX code:

```
const text = <Text>Hi, this is a message</Text>;
```

This weird-looking syntax looks kind of familiar, right? It looks like **HTML**. I'm pretty sure you have seen what HTML code looks like at least once. If you haven't, go ahead and open your favorite browser and go to something like https://reactnative.dev. Once you get there, right-click anywhere on the website and then left-click **Inspect**.

Once you've done that, you'll see lots and lots of HTML code. Randomly clicking on any of those HTML elements will take you to that specific element on the website. So, as you can see, HTML is a language that describes the way things should look, or more correctly, it semantically defines what each element is for the browser.

We don't use HTML with React/React Native, though. Instead, we use something called **JavaScript XML (JSX)**.

JSX is an extension to JavaScript that allows us to write HTML elements in JavaScript. The thing is, React Native doesn't even use the HTML part of JSX. It just uses its syntax because it makes things easier to observe and read. It is also based on React, so it's kind of obvious that it is going to be pretty similar in terms of writing code.

I feel like just by reading the preceding JSX code, we can easily understand what's going on there. It's supposed to be a text with a message stating "*Hi, this is a message.*"

We all know that "**text**" isn't the correct tag to be used in HTML because it doesn't exist. We are calling it **text** here because this is a React Native **component**.

Great! So, it's finally time to touch on components.

Discovering components

Components are – simply put – just JavaScript functions. So, we can write something like this:

```
function Element(props) {
    return <Text>Welcome to your first component</Text>;
}
```

This function is called a component because it returns a JSX element. We'll talk about **props** later, but they are really important because any component can receive a props argument inside their function.

Defining a component is easy but its usage is incredibly important. This allows us to create as many components as possible, and they can be as different as we like. This is because it clears our code out and makes things easier to organize.

Let's take a look at the code we've found in the App.js file. Try and observe the way the App() function looks. The only thing it does is return a big stack of JSX elements. In this case, we can even call this function a component and we could write it off as a JSX tag.

Expo is using this component to start your app, which means that your React Native application is just a big component that encapsulates all the other components you're going to write.

What I meant by using this component as a JSX tag is that if, for some reason, we'd like to take this component and use it in a different part of our app, we could easily just go to the file where we need it and write <App /> inside the stack. Then, everything that's inside the App() function will get rendered.

Let's try and use one of the already existent components in our only App.js file. We know <Text> is an already-defined component that's being used because we saw it work when we first tested our app.

You should already have the project and terminal open. Let's go ahead and write expo start in our terminal so that the server will start booting up.

A new window will open in your browser, just like in the previous chapter. Click **Run on…** and pick the simulator that you want to use (or use your physical device if that's easier for you). We've already discussed how to run the app, so if something seems a bit hard to understand, please go back to *Chapter 1, Introduction to React Native and Galio*, to refresh your memory.

Now that the app is alive and well on your device, we're seeing the basic screen we've already seen. Let's change it a bit by removing the text between the <Text> tags and replacing it with something else. Write your name in there; I'm going to do the same.

So, right now, we have the following line:

```
<Text>Open up App.js to start working on your app!</Text>
```

At this point, it should look something like this (but with your name instead of mine):

```
<Text>Alin Gheorghe</Text>
```

After doing that, go to the end of the line and press *Enter*. A new line will be created, so let's add something to make this starting app feel more like something personal, something that's ours.

We should add some more text describing our age and hometown. Your App() function should now look something like this:

```
export default function App() {
  return (
    <View style={styles.container}>
      <Text>Alin Gheorghe</Text>
      <Text>24, Bucharest</Text>
      <StatusBar style="auto" />
    </View>
  );
}
```

Figure 2.3 – Your recently modified code

Now, save the file you've modified (usually by hitting *Ctrl + S* or *cmd + S*) and you'll suddenly observe something cool. Once you've done this, the code automatically changed on your simulator/physical device.

This is so great, right? Usually, you'd have to restart the server, but we didn't have to do anything more than save the file we've been editing. This is called **hot reload** and is a great feature that comes packed with React Native.

Since we've added a new Text component inside our App function, you've probably guessed that we need to take this component from somewhere. You can't use a component without having it imported into your file first. So, let's learn how to do this.

Importing your first component

Now, it's time for us to learn more about importing components. Importing is great because we can grab components from anywhere and use them in our app. I mean it – you could grab a component from anywhere on the internet.

First, let's see how the Text component we've been using got into our App.js file.

If we look above the `App()` function, we will see that the first lines of code are all imports of different components. Let's take a look at those and see if they're that complicated:

```
import { StatusBar } from 'expo-status-bar';
import React from 'react';
import { StyleSheet, Text, View } from 'react-native';
```

Figure 2.4 – Imports displayed in the App.js file

It's pretty easy to read and to understand what exactly is going on here. Let's take the first line, for example. We're **importing** `StatusBar` **from** a **package** called `expo-status-bar`.

Why are we're doing that? In our `App()` function, you'll see that we've used a component called `StatusBar`.

For us to be able to use a specific component, we'll need to import it from a package or a defined path inside our project.

We can see an import from **React** but we can't find the React component anywhere inside our code; why is that? This is mostly because we need React to be able to use the React framework while creating all those components and writing JSX.

Underneath, we can see there are three different imports from a package called `react-native`. We can see `StyleSheet`, `Text`, and `View`. React Native comes packed with a lot of basic but really important implementations of native code for us to use in our React app.

We'll look at these core components in more detail in the next section, but you must understand the fact that those components were imported and then used inside our main function.

You can find *packages* online, so you could import them into your files easily by using **npm**. This is already installed with your Node.js configuration, so it's ready to use right now. We can search for packages on `https://npmjs.com` and easily install any of them with the `npm i package-name` command.

Right now, we'll focus on the components we received from `react-native`. We'll install more components in the following chapters but first, we need to learn how to use what we already have at our disposal and how we can build on top of that.

Let's start by importing some of the most important components and use them inside of our app. So, let's go to the third line in our `App.js` file. Between those brackets where we've imported `StyleSheet`, `Text`, and `View`, we'll add the `Image`, `TextInput`, and `Button` components.

Now, our line will look like this:

```
import { StyleSheet, Text, View, Image, TextInput, Button }
from 'react-native';
```

Let's try to understand what the purpose of each component is and how we can use them inside our application.

Core components

We need to understand all the basic components before we can move on. This will help us realize how to mix them so that we can create even bigger and more complex components. This will also make things easier when we're planning our app. In *Chapter 4, Your First Cross-Platform App*, we'll create a functional app for us to be proud of and our friends to look up to. The following list shows the core components:

- **View**:

    So, let's start by discussing the most important component in React Native: **View**. This component is the fundamental of all components. The `View` component is so important because you cannot build a UI without it. Acting like a container for other components, this is your best bet if you want to style something differently or arrange the layout in a specific way.

    Let's see a basic example:

    ```
    <View>
            <Text>Hi! Welcome!</Text>
    </View>
    ```

- **Text**:

    We've already used this component and it's pretty straightforward. We can use this component to display text on the screen.

    Let's see a basic example:

    ```
    <Text>This is a text</Text>
    ```

- **Image:**

    This is cool because it allows us to display an image and style it the way we want to..

    Let's see a basic example:

    ```
    <Image source={{uri: 'https://via.placeholder.com/300'}}
    />
    ```

- **StyleSheet**

  We can find an example of how this component is used by looking at our `App.js` file again. It creates a stylesheet similar to CSS but with fewer styling rules. It's really easy to use once you understand it, and we'll go further into styling once we get to our first practical challenge, where we'll create and style our very own first screen.

  Let's see a basic example:

  ```
  const styles = StyleSheet.create({
          logo: {
                  backgroundColor: '#fff',
      }
  });
  ```

- **TextInput**

  This is a component that was created for inputting text into the app using the keyboard. It is packed with all the necessary methods you'd want from an input, such as `onSubmitEditing` and `onFocus`. Don't worry – we'll use all of these when we need them.

  Let's see a basic example:

  ```
  <TextInput placeholder='email' />
  ```

- **Button**

  This component renders a basic button that handles touches.

  Let's see a basic example:

  ```
  <Button title='Press me' />
  ```

I'm pretty sure you've noticed some of these components have another word inside their tags. For example, for our `Image` component, we have the word "source," which grabs the link we're giving to know what image to display. That word is called a **prop**, and we'll learn more about them in the next chapter.

Before moving on, let's use the examples we have here for `Button` and `TextInput` in our app. We're doing this for practice and to get used to what things look like on our devices once we use these components.

Let's go and write some code displaying our age and hometown underneath our `Text` component using the examples we have for `TextInput` and `Button`. Now, the main function will look like this:

```
export default function App() {
  return (
    <View style={styles.container}>
      <Text>Alin Gheorghe</Text>
      <Text>24, Bucharest</Text>
      <StatusBar style="auto" />
      <TextInput placeholder="email" />
      <Button title="press me" />
    </View>
  );
}
```

Figure 2.5 – Your new code after importing and using the new components

Now, let's hit refresh and look at our simulator/physical device. We'll see two new things: an input that, if pressed, opens a keyboard where you can write things, and a blue button with text written in uppercase.

We haven't used the `Image` component yet as it requires styling for it to work. It needs to be told what size the image should be. We'll look at styling in more detail in the next chapter.

At this point, we've talked about all these components in a bit more detail and explained what the purpose for each one is. These are all **core components** because they deal with hardware capabilities and they need **native code** to run. By native code, we mean code written in Swift or Java for iOS or Android. Developers are building and styling components that inherit from these.

Next, we'll learn how to create components and how to organize our files so that we'll never forget where to import from.

Understanding and creating your own component

We're getting closer to our goal: creating a cross-platform mobile app. For this to become a reality, we need to learn how to create components.

First, let's create a new folder in our project's main directory and call it `components`. Here, we'll create a new file named `PersonalInformation.js`.

This folder will serve as a safe space for all our components to live in, a place where we can always import our components, just like we'd normally do with any package we'd find online.

So, we've already talked about how components are created – they're JavaScript functions that return a bunch of JSX code. However, what I haven't told you is that these components are called **functional components** and that there are different types of components out there.

Let's build our first functional component by writing all the necessary code inside our newly created file. We'll create a component whose main purpose will be to display our already written personal information on the screen.

We'll begin by writing our necessary imports. So, for this component, we know we need a `Text` component. Let's go ahead and import that. Write the following at the beginning of your file:

```
import React from 'react';
import { Text } from 'react-native';
```

We've imported React because, as I mentioned earlier in this chapter, we need it if we want to create components and use JSX. Because that's the most important and basic import, we're going to place it at the beginning of our code. After that, we imported the `Text` component from React Native.

Creating the function

Let's continue and write our functional component now, just like we learned earlier:

```
function PersonalInformation(props) {
    return <Text>some text</Text>;
}
```

Earlier, we mentioned that we need it to display the same information we did previously (our name, age, and hometown) but I haven't written anything like that. That's because we've run into our first problem.

Let's say we try to write something like this:

```
function PersonalInformation(props) {
    return (
        <Text>Alin Gheorghe</Text>
        <Text>24, Bucharest</Text>
);
}
```

Here, we'll see a bunch of red lines underneath our code. That's because JSX doesn't allow two tags to be next to each other if they're not encapsulated in a bigger tag. This is where View comes in handy. So, let's import that as well. Our second line of code will now look like this:

```
import { Text, View } from 'react-native';
```

Because we now have the View component, we can write our function with it while encapsulating our Text components, like this:

```
function PersonalInformation(props) {
    return (
        <View>
            <Text>Alin Gheorghe</Text>
            <Text>24, Bucharest</Text>
        </View>
);
}
```

With that, we've successfully created our first component. But why did we write the same thing? We already had this information in our main App.js file. We're doing this to understand why components are so cool.

Exporting and importing our component

Before we move to the main file, we'll have to be able to **import** this. We can't do this before we **export** it. Makes sense, right? Let's go ahead and add the following line to the top of the file:

```
export default PersonalInformation;
```

Now, your code should look something like this:

```
import React from 'react';
import { Text, View } from 'react-native';

function PersonalInformation(props) {
    return (
        <View>
            <Text>Alin Gheorghe</Text>
            <Text>24, Bucharest</Text>
        </View>
    );
}

export default PersonalInformation;
```

Figure 2.6 – The code we've written in our PersonalInformation.js file

If everything looks correct, save the file and move to App.js so that we can look at the most useful features of components: **reusability** and **readability**.

Now that we're in App.js, let's delete what we already have in our custom-made component – I'm talking about the Text components that are displaying our personal information. After deleting those, we can import our new component. Importing this should be easy if you've followed along so far – you just have to go underneath your last import and add another line. There, you'll import your component, like this:

```
import PersonalInformation from './components/
PersonalInformation';
```

Now, let's use this component instead of the already removed Text components we had previously. This is as easy as writing <PersonalInformation />.

Now, your code should look like this:

```
import { StatusBar } from 'expo-status-bar';
import React from 'react';
import { StyleSheet, Text, View, Image, TextInput, Button } from 'react-native';
import PersonalInformation from './components/PersonalInformation';

export default function App() {
  return (
    <View style={styles.container}>
      <PersonalInformation />
      <StatusBar style="auto" />
      <TextInput placeholder="email" />
      <Button title="press me" />
    </View>
  );
}
```

Figure 2.7 – Our code after all the modifications

Now, let's save and look at our app. As you can see, nothing has changed, but we've cleaned our code because we're only writing one line of code to get two lines of output, and that makes it a lot more natural to follow. It's a lot simpler to read because we instantly know that the Personal Information component will output personal information, and on top of that, it's really easy for us to find exactly what is of interest when we're looking for a specific part of our code.

So, if we want to go ahead and change something from our main screen – let's say we want to change our age because we're now 1 year older – you can easily see that your personal information is in a **component** called PersonalInformation that was **imported** from a folder called components. Now, all you have to do is go inside that folder, look for that specific file, and modify the text. That's easy to follow, right?

Let's create another one so that we can see how we can simplify and clean up this process even more.

Creating the Bio component

For now, let's remove the TextInput and Button components from App.js. We're not using those right now and they don't look like they have anything to do with our personal information.

After removing those from your main function, go inside our components folder and create a new file called Bio.js. This is pretty self-explanatory, but I feel like a profile should have a small biography at the top with just your name and age.

We already know that we want to import a `Text` component and create our functional component. I won't repeat the process of creating a new component; instead, I will write something personal inside the `Text` component.

> **Important Note**
>
> Don't forget that you don't need a `View` component now because we're only using a `Text` component here. The fact that we only have one JSX element means our component can easily return it without needing a parent component encapsulating it.

The new component should look like this:

```
import React from 'react';
import { Text } from 'react-native';

function Bio(props) {
    return <Text>
        I like making music and anything creative.
    </Text>;
}

export default Bio;
```

Figure 2.8 – Our new Bio component

Let's save and import it into our main file, `App.js`. As we did previously, we create a new line underneath our last import and write the following:

```
import Bio from './components/Bio';
```

Now, let's use it inside our app – I'm placing it underneath our `<PersonalInformation />` component. Save and refresh. You should now be able to see your bio underneath your age and hometown on your device.

This is great, but are we going to keep on having a new line for each component? Imagine having 30 custom components. That's going to turn into a hellish nightmare to scroll past.

Creating the main file for our components

We can easily solve this by going into the `PersonalInformation.js` file and removing the `default` keyword from the last line of our file. Do the same thing with `Bio.js`. Your last line in both files should say something like this:

```
export Component;
```

Of course, instead of `Component`, you'll have the actual name of your function, which should be `PersonalInformation` or `Bio`.

Because we've done that, we can create a new file inside our `components` folder called `index.js`. We'll create a list of all of our components here, which will allow us to import these custom components from a single line.

Inside our newly created file, `index.js`, we'll import our components and then export them. This sounds easy and somehow redundant but this is useful as it's going to make things even clearer and easier to read and follow.

After writing everything in our index file, the code inside should look like this:

```
import Bio from './Bio';
import PersonalInformation from './PersonalInformation';

export {
    Bio,
    PersonalInformation
};
```

Figure 2.9 – The index.js file with all the code written inside it

Now that we have this file that stores all of our newly created custom components, let's go into our `App.js` file and rewrite our imports the way they should be written.

Refactoring our main code

Here, we must delete our first two custom component imports and write the following code:

```
import { PersonalInformation, Bio } from './components';
```

That's the only change we're making. Pretty easy, right? And it looks so much better and more organized.

Now, let's remove the unused components, such as `Text` and `Image`, and save our file. After making all these modifications, your `App.js` file will look like this:

```jsx
import { StatusBar } from 'expo-status-bar';
import React from 'react';
import { StyleSheet, View } from 'react-native';
import { PersonalInformation, Bio } from './components';

export default function App() {
  return (
    <View style={styles.container}>
      <PersonalInformation />
      <Bio />
      <StatusBar style="auto" />
    </View>
  );
}
```

Figure 2.10 – Our final code for this chapter

Yay! We've finished creating two new components for our app while also organizing the code in such a manner that any programmer would be proud of us. I'm not a believer in homework but I do believe in the power of exercise. Now, it's your turn. Create as many components as you can think of. Don't stop at simple text-based components; try and use more of the core components React Native has at its disposal. Don't be afraid of getting something wrong – that is the best way to learn: trial and error.

Summary

In this chapter, we started learning about our Expo's basic file structure and how all those files are connected, how `App.js` is the main entry point into our application, and which function is getting called at startup. After that, we delved into the main concepts of JSX, explaining and comparing JSX to other markup languages and understanding that JSX is more of an extension of JavaScript.

We left the theory aside and started importing our first component while talking about npm and how we will use it in the future when creating more complex applications. We imported the core components of React Native and explained them all. Using them felt comfortable and pretty easy, so we figured, why not create a component? After we created a component, we learned more about file structure and how to index all of our components into a single file, which helped us clean our code even more.

In the next chapter, we'll study the correct mindset of a React/React Native developer and understand how to think in React. This is going to help us greatly because it will save us time when we're starting a new project. If the planning is correct from the start, we won't have any problems building the project.

3
The Correct Mindset

I think I'm right in saying that we've learned quite a lot together, and I hope you're excited to keep up the learning process. In the previous chapter, we learned more about how a React Native project works and what role each file or folder has. After that, we started learning about **JavaScript XML (JSX)** and how to use it, and we've actually imported our first component. Learning about the core components you're going to use every time a new project is created set us on the path of understanding and creating our own component.

This chapter will focus mostly on the React architecture and how it makes us think in a certain way after we spend some time with the framework. For starters, we'll begin with the main idea about how people start a React application—or, in our case, a React Native application, and we'll easily transition into the more advanced concepts of React, such as props.

By grasping the concept of props, we'll be able to add the next level of complexity to our applications. This will allow us to create even cooler components, unlocking more powers of React. You'll find yourself using props in almost every component you create.

After that, it's time to learn about rendering lists and how to use those to change the information inside our components. This sounds pretty neat, right? We'll be able to have different information shown based on whatever calculations we want to do inside the component or however many items we need to showcase.

Completing this chapter will teach you great ways of thinking as a React developer. This will serve as a good time saver for when you first start any project, and it's really important to understand how to correctly structure our files and code.

We'll realize how programmers are reusing their code in such a way that you'll keep repeating to your family: "Write once, use everywhere!"

This chapter will cover the following topics:

- Thinking in React
- Always build the static version first
- Props and how to use them

Technical requirements

You can check out this chapter's code by going to GitHub at `https://github.com/PacktPublishing/Lightning-Fast-Mobile-App-Development-with-Galio`. You'll find a folder called `Chapter 03` that contains all the code we've written inside this chapter. In order to use that project, please follow the instructions found in the `README.md` file.

Thinking in React

Let's not forget the fact that Facebook created React for their own projects and it's actually used in almost any type of website (or mobile app with React Native), so it has big scalability features. If Facebook can use it in their platform, we can surely use it inside our apps.

For us to take full advantage of this framework, we need to start thinking in React. When I first started my programming journey, the idea of a framework seemed kind of alien to me. I didn't understand the fact that it's called a framework because it comes packed with a specific workflow. Well, that's not the only reason why it's called a framework—it's also because it comes packed with tons of features and methods to make our work easier.

Let's imagine we're working with our friends on an app idea that we're going to call *PiggyBank* just for fun. The idea is that we need to always keep track of all the transactions we're making with our credit card. So, this basically means we'll have a card keeping track of all our transactions somewhere in our app.

I've designed the following card in Adobe XD, which I think will help us visualize things better:

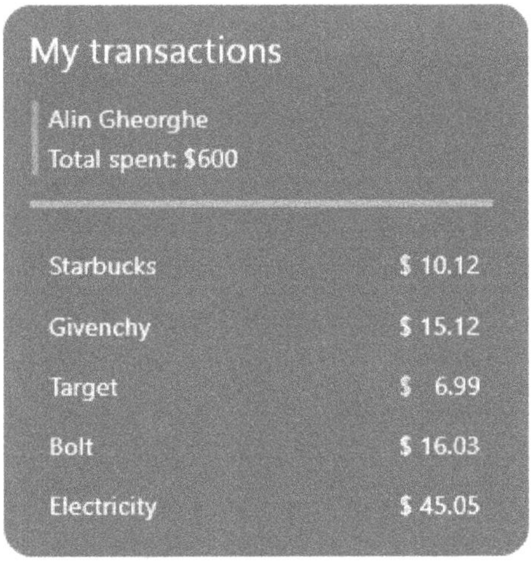

Figure 3.1 – Card component displaying our transactions

So, our friend came up with this cool card design and he's asking us to implement it in the mobile app. Easy, right? We've seen how everything can be coded from top to bottom just by using JSX code; on top of that, there's so much text on this card that it makes things even easier for us. You might even think we don't need any custom components or we might only need one.

Well, that's not entirely true. This is the moment where our React knowledge shines through and it helps us divide everything into components for much easier and cleaner code. But the question remains… how do you know where to draw the rectangles, and how do we delimitate the components? One technique that React recommends is the **single-responsibility principle (SRP)**.

> **Tip**
> The SRP is a programming principle that dictates the fact that every class in a program we're writing should have responsibility over a single part of that program's functionality. This is part of **SOLID**, which is an acronym for five design principles for software engineers to create more maintainable, flexible, and understandable code.

Using this principle, we should now be able to divide the card into components. Let's take this card and draw rectangles on it for each and every component we encounter, as follows:

Figure 3.2 – Rectangles drawn to divide components

So, we've extracted the following components:

- `TransactionCard` (*red*)—Contains the full card and all of its elements
- `TransactionCardHeader` (*green*)—Represents the upper part of the card, the part with the name and total money spent
- `TransactionCardList` (*yellow*)—Contains a list of items
- `TransactionItem` (*pink*)—A single item displaying the transaction and the price

As you can see, we've successfully divided this card into four different components that will have to somehow talk to each other and at the end have this one purpose of showing information regarding our transactions.

Each of them has only one single purpose, so that checks the SRP we've been talking about. Let's code it, without styling yet—we'll do that in the following chapter. Always build the static version first.

First of all, we need to understand that the easiest way we can start working on an application is by building static pages and then separating everything and building the logic. For that, we'll have to basically copy everything we see in the image, even the text— that's why we call it a static version.

Let's start by creating a new project by opening up the terminal, moving to our project's folder, and writing the following command:

```
expo init TransactionCard
```

I've picked `TransactionCard` as my project name, but don't forget that you can name it whatever you want. Next, we'll be choosing the blank managed workflow template and will wait for it to start initializing our project. Once that's done, let's open up our **integrated development environment** (**IDE**)/text editor of choice and look over the project.

I'll open up `App.js` and delete the `StatusBar` import (not really important for this exercise) and everything from the inside of our `View` component in our `App` function.

Let's decide what types of components we need. It's pretty easy to see that this card only needs a `View` component and a `Text` component. Luckily for us, we already have those imported in the file, so let's use them to build our static version of the card.

We'll start by looking at our design and try to divide everything into containers. As we can see, this card has two parts: the upper part is the header with general information, and the lower part is full of all of our recent transactions.

So, let's write the code for those parts, but first, we'll focus on the upper side of our card, as follows:

```
export default function App() {
  return (
    <View style={styles.container}>
      <View>
        <View>
          <Text>My transactions</Text>
          <Text>Alin Gheorghe</Text>
          <Text>Total spent: $600</Text>
        </View>
        <View>

        </View>
      </View>
    </View>
  );
}
```

Figure 3.3 – Upper side of our card

We see a lot of `View` components here, but why is that? As we discussed in the last chapter, the `View` component is usually used for layout design and grouping elements together. So, we've grouped the two parts inside of it and then wrote code for the header. If we save and open our simulators, we should be able to see the text we've just written.

> **Important note**
> Don't forget that if you use the Android simulator, you first need to open up Android Studio and then go to the **Android Virtual Device (AVD)** Manager. Run your simulator and afterward start the app from the Expo dashboard.

Let's think about our transactions now. We have two pieces of text—to the left is the name of the company from which we bought whatever we bought, and to the right, we can see the price of the transaction. How can we do that? As far as we have seen until now, the elements we've printed out to the screen were always aligned in a columnar style from top to bottom.

Well, we need to style it for that to work, but the important point here is that we need to understand the fact that those two components are somehow connected, so we'd have to put those elements in the same `View` component.

Let's do that now and get our components ready for writing all these transactions. First, try to do it on your own and then see if you got the same code as I did.

I'll start by creating a different `View` component for each line and then add the text inside it, like this:

```
return (
  <View style={styles.container}>
    <View>
      <Text>My transactions</Text>
      <Text>Alin Gheorghe</Text>
      <Text>Total spent: $600</Text>
    </View>
    <View>
      <View>
        <Text>Starbucks</Text>
        <Text>$ 10.12</Text>
      </View>
      <View>
        <Text>Givenchy</Text>
        <Text>$ 15.12</Text>
      </View>
      <View>
        <Text>Target</Text>
        <Text>$ 6.99</Text>
      </View>
      <View>
        <Text>Bolt</Text>
        <Text>$ 16.03</Text>
      </View>
    </View>
  </View>
);
```

Figure 3.4 – The rest of the static code

Let's open up our app and take a peek at how everything looks. Right now, it's just a bunch of text in the form of a column displaying all the information we have in our design. It doesn't look like a card, though, but we can definitely see a resemblance to our design in terms of having the same information in the same order. So, we've now created the most basic version of our card.

The next step is to finally break down this big tree into smaller components. By doing this, we'll make the code easier to read and understand and also more modulated, which basically means we could use the same component in a different card or whatever we need it for at that moment.

Breaking down our code

So, remember we've been dividing the design into four different components? Let's create a `components` folder and create four different files for each component, as follows:

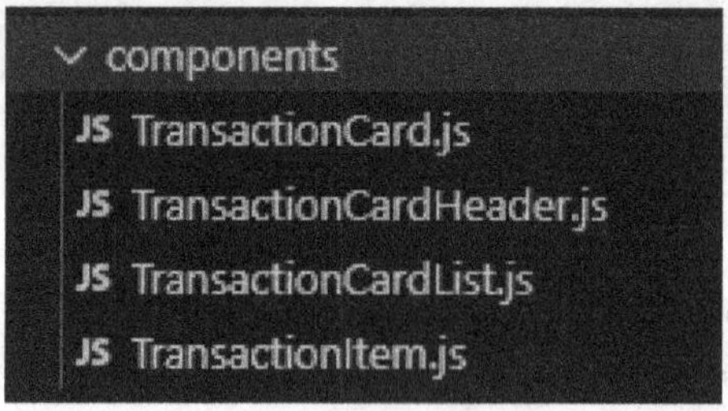

Figure 3.5 – Our folder with all the files created

Now, it's time to start coding each one. So, we know that the big card—or the first component—should be able to be split into two parts, the `List` and the `Header` components. Let's copy all of our code from `App.js` into `TransactionCard`, as this one is the main component. By all the code, I mean only what's inside the first `View` component.

After creating our function, we paste all the code inside it. Let's export the component and take a look at it, as follows:

```
const TransactionCard = () => {
    return (
    <View>
        <View>
            <Text>My transactions</Text>
            <Text>Alin Gheorghe</Text>
            <Text>Total spent: $600</Text>
        </View>
        <View>
            <View>
                <Text>Starbucks</Text>
                <Text>$ 10.12</Text>
            </View>
            <View>
                <Text>Givenchy</Text>
                <Text>$ 15.12</Text>
            </View>
            <View>
                <Text>Target</Text>
                <Text>$ 6.99</Text>
            </View>
            <View>
                <Text>Bolt</Text>
                <Text>$ 16.03</Text>
            </View>
        </View>
    </View>
    );
};

export default TransactionCard;
```

Figure 3.6 – TransactionCard component written as an arrow function

We've written this component as an arrow function because it's easier to write—at least, in my opinion—but honestly, you can write it even as a class if you want. As a rule of thumb, you usually use `class` only if there's state involved, but state is something we need to go into in more depth in the later chapters.

So, we have all the code here and we've exported our function. All good—now, the next step is to go even deeper with our division. Let's take the header side of our component and move it to its specific file.

After we've finished copying the code into our `TransactionCardHeader` component, let's import the component into our `TransactionCard` component and use it instead of our copied code. We should do the same thing with the second part of our card, and that is the `TransactionCardList` component. Let's do it and see what everything looks like. Here's the result:

```
import { View } from 'react-native';
import TransactionCardHeader from './TransactionCardHeader';
import TransactionCardList from './TransactionCardList';


const TransactionCard = () => {
    return (
    <View>
        <TransactionCardHeader />
        <TransactionCardList />
    </View>
    );
};
```

Figure 3.7 – Our newly created TransactionCard component

OK—so, this looks much, much cleaner. If we import this component into our `App.js` file, everything should look identical to how it used to look before we started making all these changes to our code.

> **Tip**
>
> Don't forget that we always need to run `import React from 'react';` so that we can use all of our components. For a component to run, it needs to know it's a component and not just random writing in a file. That import helps our code identify which objects are React objects and how to render everything.

Everything works, right? If you've encountered any issues, stop for 2 seconds before moving further and check everything we've been doing until now; maybe you got something misspelled or you forgot some exports somewhere in your files.

If everything's fine, let's get inside our `TransactionItem` component. Well, as the name suggests, this is one single item, so what does that mean? As we can see in our `TransactionCardList` component, we do have several different items. Are we going to create a different component for each one of them?

Not really—we're actually going to create a single component that changes the information displayed based on whatever information it receives as input. This sounds pretty cool, right? Well, that input is called a prop, and each component gets a set of props when it renders by default, but it can also receive custom props created by us. Let's dive into props and learn how to use them in the context of our card.

Props and how to use them

So, what exactly are props? Up to now, we've only used normal tags to identify our components such as `TransactionCardHeader`. However, as we saw earlier when we presented different components, these ones could also have **props** that are used to pass down information from the bigger component (**parent**) to a smaller component (**children**).

Let's go into `TransactionCardList` and look at our code. As far as we can see, there's a lot of code repeating itself in terms of components being used. So, we can see this pattern emerging from inside our main `<View  />` tag:

```
const TransactionCardList = () => {
    return (
    <View>
        <View>
            <Text>Starbucks</Text>
            <Text>$ 10.12</Text>
        </View>
        <View>
            <Text>Givenchy</Text>
            <Text>$ 15.12</Text>
        </View>
        <View>
            <Text>Target</Text>
            <Text>$ 6.99</Text>
        </View>
        <View>
            <Text>Bolt</Text>
            <Text>$ 16.03</Text>
        </View>
    </View>
    );
};
```

Figure 3.8 – TransactionCardList component ready to be divided into smaller components

The pattern is pretty easy to see—we have four identically written pieces of code but with different information written inside of it. We basically have four instances of `View` components with two `Text` components inside of them. Seeing how this repeats, we can clearly realize that we could write an external component for this specific case.

Let's begin by writing a static component inside our `TransactionItem` component and see how we can implement it inside our list.

Just go ahead and write one of these pieces of pattern; we only need one—otherwise, we'll kind of defeat the purpose of a singular item. This is what the code should look like:

```
import React from 'react';
import { View, Text } from 'react-native';

const TransactionItem = () => {
    return (
        <View>
            <Text>Starbucks</Text>
            <Text>$ 10.12</Text>
        </View>
    );
};

export default TransactionItem;
```

Figure 3.9 – Static version of TransactionItem

Now, let's use this component instead of all our pieces of pattern we've used inside our list. After importing the file and replacing everything with four or five instances of `TransactionItem`, we can see that the data is now `Starbucks` and `$ 10.12` everywhere. Repeating itself like crazy is not really how a great mobile app is designed, right?

Now, how can we change that? How can we have different pieces of information displayed by our component? By using props. Let me change the `TransactionItem` component and see how props need to be implemented. This is what the code should look like:

```
const TransactionItem = (props) => {
    return (
        <View>
            <Text>{props.name}</Text>
            <Text>$ {props.price}</Text>
        </View>
    );
};
```

Figure 3.10 – Implementing props into our component

Right now, your `TransactionCardList` component contains multiple instances of `TransactionItem`. If you save right now there's nothing showing up for these components besides the $ sign. Why is that?

This all happens because our component has nothing stored in those variables. For us to actually get something displayed on the screen, we'd have to send some information over to our `TransactionItem` component from `TransactionCardList`. Let's move inside of it and use our newly updated components to display the correct information on our phone.

Inside the `TransactionCardList` component, find our components and add the following props to each component, like this:

```
<TransactionItem name={"Starbucks"} price={10.12} />
```

After we've added props to all of our components, the next step is to save. We'll see how our simulator refreshes automatically—and congratulations! We've successfully sent information from one **parent** component to our **child** component.

> **Tip**
>
> All the information being sent from one component to another will formally be inside curly braces, as we've seen by writing the number for the price prop. Even though strings could still be placed inside curly braces, they're not mandatory for sending information, so you could even write `name="Mircea"`.

Now, let's try to understand our code a little bit. So, what's going on, really, inside our app?

When the app first runs, it goes straight to `App.js` and starts rendering all the components first written there. For that, that'll be our `TransactionCard` component.

React sees that our component actually has two different components inside it and begins rendering the next components. Now, one of these components is actually our `TransactionCardList` component, which contains all our `TransactionItem` components.

Because the first component contains another component, we call the first one a **parent** and the second one a **child** to the first one. So, if `TransactionItem` is a **child** to `TransactionCardList`, try to figure out what `TransactionCard` is to `TransactionCardHeader`. Ready? `TransactionCard` is the **parent** of `TransactionCardHeader` because it contains the other component.

Now, when React reaches `TransactionCardList`, it'll send some information to each `TransactionItem` component via the **props**. The information being sent is a JavaScript object that looks like this: `{name: 'Starbucks', price=10.12}`.

That's why we can use props as an argument for our function in `TransactionItem` and then access the keys of our object with a dot, like this: `props.name`. You must be wondering how React knows how to handle all these processes because a more complex app might have hundreds of components nested into each other while sending props to one another at the first render.

The thing is, React first renders everything that's at the surface, and when all the information is being sent from one parent component to a child, then it renders that information.

Now, just to make our component even more usable and reusable, we'd have to make the number of items in our list a little more variable. For a bigger app, we'd have to ask ourselves questions such as these: "What if the number of transactions a user makes is bigger or smaller than five?"; "What if I'll need more cards for future screens but with the same design? How can I reuse this card component?"

Using the map function to dynamically change the number of components

Well, let's first see how we can output as many `TransactionItem` components as we can. We'll go inside our `TransactionCardList` component and create a constant array of objects, outside of our function, called `transactions`. This variable will contain all the information needed for our items. Let's see what this looks like, as follows:

```
const transactions = [
    {name: "Starbucks", price: 10.12},
    {name: "Givenchy", price: 15.12},
    {name: "Target", price: 6.99},
    {name: "Bolt", price: 16.03},
    {name: "Electricity", price: 45.05}
];
```

Figure 3.11 – transactions variable

Once we have this variable with all the information that we need, we could **map** our array and output a different component for each item. This might sound a bit confusing if you're not really comfortable with JavaScript, but trust me, it's really easy. Let's delete everything from inside the `<View />` component and replace it with the map function, as follows:

```
const TransactionCardList = () => {
    return (
    <View>
        {transactions.map((item, key) => {
            return <TransactionItem name={item.name} price={item.price} key={key}/>
        })}
    </View>
    );
};
```

Figure 3.12 – map function used inside TransactionCardList component

OK—so, this might look a bit strange. Don't worry—it's actually really easy. So, we've used the map function on our `transactions` array. This map function goes over each element of our array and uses the function inside its argument to output *something*. That something is where you come in and make use of this cool function.

> **Important note**
>
> All external code used inside JSX **must** be put between curly braces so that React can understand that we're doing an operation that could result in other elements being outputted for rendering.

Basically, because of the map function, we're taking the first item of our array— `{name: "Starbucks", price: 10.12}`—outputting a `TransactionItem` component, and passing as props the values we've had in our array. But we also see the **key prop** and we both know we haven't used the key prop inside our component. Every list needs a key for each child so that React can keep track of them and avoid excessive re-rendering. This is one of React's rules that we need to understand whenever we're using lists such as these.

But we've said we'd go even further, right? We need to use this card component multiple times if we need to. Seeing how `transactions` is just a random variable sitting in our `TransactionCardList` component, maybe we could send that as a **prop**?

Let's add the `props` keyword inside our function's argument and change from `transactions.map` to `props.transactions.map`. If we save now, we get an error—our component expects a prop called `transactions` to come in but there's nothing sending it.

We'd have to send this from our parent component—namely, `TransactionCard`. But nonetheless, this doesn't really change the fact that we still can't use the card properly, so maybe we need to add this prop to even our `TransactionCard` component.

Let's copy our `transactions` variable and move it inside our `App.js` file. After that, let's add the `transactions` prop to our `TransactionCard` component, like this: `<TransactionCard transactions={transactions} />`.

Now, we'll have to go to our `TransactionCard` component and enable it to take this prop and send it even further to our `TransactionCardList` component. Our component now needs to look something like this:

```
const TransactionCard = (props) => {
    return (
        <View>
            <TransactionCardHeader />
            <TransactionCardList transactions={props.transactions}/>
        </View>
    );
};
```

Figure 3.13 – Our newly created version of the TransactionCard component

So, we've been sending this information from the `App.js` file all the way to our `TransactionItem` component where we finally display the information. How did this help us? Well, right now, we can have multiple instances of this card with different transactions, or we can even add or lower the number of transactions based on the constant that we now have declared inside our `App.js` file. We can use different variables altogether; we can have a different array called `biggerTransactions` and pass it to another component. Maybe this one will display the biggest transactions you've done.

The important thing here is the fact that we now don't have to touch our card component at all and we can still use it while displaying different information. This is much easier than creating different files for each piece of information we need, or maybe at one point you need to change specific information and you start browsing every file looking for that specific thing. You now don't have to do that—just go into your main file and change all the information from there.

Let's do some homework. You'll find the answer in our GitHub repository, in the `Chapter 03` folder. Having the same name all the time on our card could become boring. Make this easier by allowing yourself to use multiple instances of the same card component but for different users. After finishing this, go check out the code and compare it to mine. See if you did the same thing!

Summary

In this chapter, we've gone even deeper with React Native. We've learned so much about new concepts such as props and the SRP. We should be able to start thinking using the React methodology based on props for now, and, later, even state. But understanding all this is a great step toward you becoming a real React Native developer.

You should feel even more comfortable regarding the way props are handled and how we can use this special feature of React called components for our benefit, for reusability and cleaner code. There's no such thing as your code being too clean, but at the same time, keep in mind the fact that sometimes there's no need for multiple layers of props. Maybe your component needs only one layer or no props at all. Only use this feature when you feel it might make your work easier.

We've also created a list for the first time and learned that each item of a list needs a key, a key that could sometimes be even our array's index, but there's always a unique key being sent to each of our items.

At the end of this chapter, we finished with a little bit of homework and a lot of hope for the next chapter, where we'll finally create our first small application to show our friends.

4
Your First Cross-Platform App

We started by learning how to set up a React Native development environment. After that, we went ahead and started learning about JSX, components, and props. We've already learned so much that we should be quite confident going forward. But if you still feel like something is missing, then you're right. We haven't styled anything, and we still haven't constructed a real screen.

This chapter will revolve around an app idea I had some time ago that constantly tracks your gaming history. We're not going to start discussing servers and databases as they are out of our learning scope, especially because we have bigger, more essential things to learn about. We'll start by detailing all the information about our app while using everything we've learned in the previous chapters.

After that, we'll take start creating the static version of our app so that you can understand how your brain needs to think before creating an app. Having learned all the principles in the previous chapters will help us understand our first real assignment more easily, so if there's anything you still aren't sure about, go back to the previous chapters and see where exactly you feel like things could be improved.

The next step is to learn about styling. We'll go in-depth regarding styling and how that works with React Native. We'll understand what flex is and how to use it inside our app while also figuring out tricks we can use to make our development easier.

After styling the app, we'll refactor our code while keeping everything we've built so far intact. This is where Galio will come in and help us realize how useful it is to have already-built components at our disposal. We'll learn how to use one of the most important components to build the layout without worrying about creating different styles for our containers.

After this, we will install the app on our phones. This is a single-screen app, so we'll be using our phones for testing purposes only. We'll also learn some basics techniques that we can use to make sure our app runs smoothly on all screen sizes.

Everything seems straightforward and pretty easy, right? Let's head right in and start building our app. The following topics will be covered in this chapter:

- Building our first app

- Creating your first screen

- Let's style it!

- The superhero, Galio

- Let's install it on our phone

Technical requirements

You can check out this chapter's code by going to GitHub at `https://github.com/PacktPublishing/Lightning-Fast-Mobile-App-Development-with-Galio`. You'll find a folder called `Chapter 04` that contains all the code we've written inside this chapter. To use that project, please follow the instructions in the `README.md` file.

Building our first app

Let's start discussing the main idea of our app and how we're going to start building it. We'll call this app **MGA**, which is short for **My Gaming History**. Pretty clever, right? It will only have one screen, and it will act as the welcoming screen once the user is logged in. We'll pretend that the user has already logged into our app, so we'll only code that main screen without looking at authorization, which is a more advanced concept.

By having a clear understanding of what our screen needs to look like and describing our component's purpose, we're building a clear path to our development. In the end, if we weren't doing all this preparation, we'd get stuck during our programming, and we don't want that.

We should start by looking at our design, identifying the main purpose of it, and how to start sectioning our screen into components:

Figure 4.1 – My Gaming History's main screen

It looks great, right? Well, it should because this time, we're going to fully implement everything in this screen, even the colors and element positioning. After all, this is our first fully created screen.

Let's try to think about how we could separate everything into smaller sections, which is one of the most important steps in UI creation. Remember that this is mandatory because if we were to just code everything without even trying to have some sort of strategy in mind, we would run into some issues.

We're going to use squares to easily identify each element on our screen, so let's take a look at that:

Figure 4.2 – Component division

I've sectioned the entire screen here, color-coded so that you can see them better:

- **Home** (*red*): Our container component, also known as our screen.
- **WelcomeHeader** (*blue*): This will contain all the basic information regarding the user, such as their name, level, and profile picture.
- **MostPlayedGame** (*blue*): This will be a container that will receive information regarding the most played game but also a picture.
- **LastPlayedGameList** (*blue*): This contains a list of items.
- **PlayedGameItem** (*green*): This is a single item displaying the most played games and the times spent on each one.

As we can see, we're using the same color for three different components. Why is that? Because those three components are equally important in our main bigger component called Home. They're all sitting at the same level in our component tree. Even though the Home component is a screen, it is defined the same way as a component, and you'll see what I mean by that when we start coding.

Now that we have divided our components, we're ready to move on and start coding our app.

Creating your first screen

Once the development plan has been completed and we know where each component needs to go and what our application is going to look like, we're ready to create a new project. This project will set the first stones for this creative path that we've taken toward being a React-Native developer.

Let's begin by creating a new project:

1. Go to your Terminal in your preferred directory and run the following command:

    ```
    expo init mga
    ```

2. Choose the blank template for *Managed Workflow* and open the project folder.

3. As we mentioned previously, we'll have five different components, one of which will be the screen itself. So, let's create two different directories called `screens` and `components`. This will make it easier to organize when we have several different screens.

    It's a good rule of thumb to always have a basic structure in your head when you start coding because you never know when you might want to add more and more to your app.

4. Inside our `screens` folder, let's create a file named `Home.js`. This will be our main screen, so we will begin by writing the most essential code for a component. This is just the boilerplate for a functional component. Remember how those were created? We did this in *Chapter 2, Basics of React Native*. Now, try to do it by yourself and come back here once you've managed to do it:

```
import React from 'react';
import { View } from 'react-native';

const Home = () => {
    return (
        <View />
    );
}

export default Home;
```

Figure 4.3 – Basic Home screen with nothing to render but a View component

5. Once you've done that, we must go to our main file, `App.js`.

    Here, we'll start by removing everything that's not necessary for our main file. We won't need all the styling, the `StatusBar` import, or even any components that have been imported from React Native.

6. After deleting everything, we can import our component right after the `React` import and place it inside our main `App` function.

    So, our new component should now look something like this:

```
import React from 'react';
import Home from './screens/Home.js';

export default function App() {
  return (
    <>
      <Home />
    </>
  );
}
```

Figure 4.4 – The App.js file after removing all the unnecessary code

You might be wondering, "Well, what's with that weird <> syntax?". That's the short syntax for a **Fragment**, which is a React feature. This is used so you don't add more unnecessary nodes to your component tree. We could have used a `<View />` component instead, as we saw in our earlier examples, but by using a **Fragment**, we're creating a wrapper for our components without an unnecessary wrap component as we won't be styling anything in our main file.

If this still creates some problems, you can easily wrap your `<Home />` component into a `<View />` component.

Now that we're here, let's look at our `components` folder and create all the necessary files we'll be working with.

7. Create four new files called `WelcomeHeader.js`, `MostPlayedGame.js`, `LastPlayedGameList.js`, and `PlayedGameItem.js`.

8. Let's do the same thing we did with `Home.js` for each of our newly created files. You could even copy the code from the `Home.js` file and then paste it into each file; just don't forget to change the names from **Home** to whatever your component is called.

Now that we have initialized all our files, we're ready to start moving on with our code. We should see some sort of similarities between these components and the previous chapter's components. It's almost the same thing, so you should have an idea about how we could move forward.

We'll start with `WelcomeHeader.js` and then look at each of our files. If you were to start up your app right now, you'd see a blank white screen. We'll ignore that for now and just sketch our app with some basic static code so that we have some sort of a basis for when we begin styling.

Open your file so that we can start adding some new elements. What can we observe from the design that we might need inside our component? Well, first of all, there's a lot of **text**, but we also need a **profile picture** (that circle on the right-hand side of the component). Knowing this, we can now start importing the components needed for this, so go ahead and edit the second line where we're importing the `View` component so that it looks something like this:

```
import { View, Text, Image } from 'react-native';
```

Remember when we said we should group components if they're on the same line? This will make things easier when we begin styling as those components are on the same horizontal line.

So, I've started by adding another `View` inside our main `View` component. After that, I'll add the components that are sitting on the same line: our welcome *message* and our *profile picture*. Beneath this `View` component, we'll add another `Text` component that will render our `Level`:

```
import React from 'react';
import { View, Text, Image } from 'react-native';

const WelcomeHeader = () => {
    return (
        <View>
            <View>
                <Text>Hello,{'\n'}Alin Gheorghe!</Text>
                <Image source={{uri: 'https://via.placeholder.com/65'}} />
            </View>
            <Text>Level 3</Text>
        </View>
    );
}

export default WelcomeHeader;
```

Figure 4.5 – Our static version of WelcomeHeader

In *Chapter 2, Basics of React Native*, we discussed how an `Image` needs a source to work. That's why we've used the `source` prop and passed a placeholder image link. It's easier to use placeholders as we don't need to waste time searching for images when our main purpose is to just code a static version for now.

Let's move on and start coding our next component: `MostPlayedGame`. As we can see, here, we need the same things as we did in our previous component. So, let's import everything and use it inside our component. Once you've done that, we will use our components to display all the information. Now, your code should look like this:

```
import React from 'react';
import { View, Text, Image } from 'react-native';

const MostPlayedGame = () => {
    return (
        <View>
            <Image source={{uri: 'https://via.placeholder.com/300'}} />
            <Text>Most played game: League of Legends</Text>
        </View>
    );
}

export default MostPlayedGame;
```

Figure 4.6 – Our static version of MostPlayedGame

I've written 300 instead of 75 in our placeholder link because that changes the width of the image. But beside that, this was pretty easy to understand.

At this point, we'll notice something really interesting. We have a list that follows the same pattern we're used to. It's a list of items, and each of those items renders a game we've played and how much we've been playing it. We could copy the same pattern we've used previously and it would work exceptionally well:

```
import React from 'react';
import { View, Text } from 'react-native';

const PlayedGameItem = (props) => {
    return (
        <View>
            <Text>{props.game}</Text>
            <Text>{props.hours}h</Text>
        </View>
    );
}

export default PlayedGameItem;
```

Figure 4.7 – The PlayedGameItem component

I'm sure you remember how easy it was to pass `props` from a parent component to a child one. We shouldn't be wasting any time if we already know how certain components should be coded. Now, it's time to create the list, just like we did the last time, but now, we have another element inside of it, a `Text` component acting as a header for our component:

```
import React from 'react';
import { View, Text } from 'react-native';
import PlayedGameItem from './PlayedGameItem';

const LastPlayedGameList = (props) => {
    return (
        <View>
            <Text>Last played games:</Text>
            {props.games.map((item, key) => {
                return  <PlayedGameItem game={item.game} hours={item.time} key={key} />;
            })}
        </View>
    );
}

export default LastPlayedGameList;
```

Figure 4.8 – Our finished LastPlayedGameList component

We've moved pretty fast but that's because we've already been through this, so you should understand what's happening here. The problem with our code right now is that we're not sending any information to our items. We don't have that array that our `map` function needs to run. As you can see, the array comes from `props`, so our `LastPlayedGameList` component is expecting a **prop** called `games` with an array so that it can start rendering our list of games.

Let's move inside our **Home** screen component and set everything up. First, we'll begin by importing all the components needed for our screen. We only need three out of these four components because one of them is `PlayedGameItem`, which is already being used and rendered by our `LastPlayedGameList` component. Importing them is easy, as shown here:

```
import WelcomeHeader from '../components/WelcomeHeader';
import MostPlayedGame from '../components/MostPlayedGame';
import LastPlayedGameList from '../components/
LastPlayedGameList';
```

After importing everything we need, it's time to place the components in the order they'll appear on the screen, inside our main `View` tag:

```
const Home = () => {
    return (
        <View>
            <WelcomeHeader />
            <MostPlayedGame />
            <LastPlayedGameList games={games}/>
        </View>
    );
}
```

Figure 4.9 – Our Home component with the rest of the components inside it

As you can see, I've already passed the `games` array we need for our list, above our component. Let's create an array so that we have something to pass over to our `LastPlayedGameList`.

First, try it out for yourself – remember that we need an **array of objects** with the `game` and `time` keys. Once you've tried this for yourself, come back here and take a look at the following code:

```
const games = [
    {game: "Bioshock", time: 2},
    {game: "Call of Duty", time: 5},
    {game: "GTA 4", time: 1},
    {game: "Need for Speed", time: 3},
    {game: "Among us", time: 1},
];
```

Figure 4.10 – The games object ready to be sent over to our list component

That wasn't that hard, right? Here, we've coded our entire static screen. I'm pretty sure you should be able to see something popping up on the screen if you were to go back to your simulator. If there aren't any errors, then we should be able to move on. If you did encounter any errors on the screen or you still can't see anything, try to reread everything and make sure you haven't missed a word. I'd say 70% of errors are thrown out in the development stage because we're usually missing some characters in our variables (don't quote me on that, it's just personal experience). JavaScript, being a **loosely typed language**, means you don't have to specify what type of information will be stored in a variable, so we don't have to worry about defining variables incorrectly as much as a **Java** or **C#** developer has to but at the same time, a variable needs to have the same name wherever it's used.

Now, let's start making it pretty.

Let's style it!

Before we start styling our app, we should understand how styling works in React Native. If you do have any prior experience with React, you'd know that styling is done via CSS. However, in React Native, we can't use CSS, so everything is done through the StyleSheet.

A StyleSheet was created by the React Native team. Here, you have similar rules to CSS but everything is done through JavaScript.

When passing these styling objects over to our components, we do so via a prop called `style`. Let's start by directly creating some styles for our **Home** screen.

There are two ways we can pass these objects over to our components – we could write them directly in our component or pass them via a new instance of the StyleSheet. Let's write it in-line and change our background color for the screen. By going over to our Home.js file, we can add the `style` prop to the our `<View />` component, which wraps the rest of our components:

```
const Home = () => {
    return (
        <View style={{flex: 1, backgroundColor: '#F8EDEB'}}>
            <WelcomeHeader />
            <MostPlayedGame />
            <LastPlayedGameList games={games}/>
        </View>
    );
}
```

Figure 4.11 – Adding in-line styling to our component

After adding this and saving the file, you should be able to see how the whole background color changes to that hex color. Now, our background color is the same as the design image's. This is pretty cool, right? It's also easy to read as this is essentially CSS but written a bit differently.

If we were to write CSS, we'd be saying, for example, `background-color: 'red'`, but because everything is JavaScript in React Native, we can't write variables or object keys with the dash between characters, so we're using camel case.

But there's an issue regarding the in-line styling; we could easily have thousands of styles and in that case, we're going to forget where some things are or how to change certain things in our app. That's why we should try and use a cleaner way of writing styles.

Let's delete our in-line styling and start by changing the import by adding `StyleSheet` next to `View`, like this:

```
import { View, StyleSheet } from 'react-native';
```

Now that we have imported `StyleSheet`, we're ready to create some styles. To do that, we will use the `.create()` method. This method will return an object with all the necessary styling information:

```
const styles = StyleSheet.create({
    container: {
        flex: 1,
        backgroundColor: '#F8EDEB'
    }
});
```

Figure 4.12 – The styles object

Right now, we can go back to our `<View />` component and inject the styling into our style prop by using `style={styles.container}`. Now, everything should look identical to how it looked when we had our inline styling. I'd recommend using the `.create()` method to add styles as it's much cleaner and easier to read.

Now, you might have some questions regarding `flex`. I mean, you've seen it there, but you haven't realized what that property is doing yet. Those questions should extend to "Can I use all CSS rules inside React Native just by writing them in camel case?"

The thing is that CSS has two options for the layout: **Grid** and **flexbox**. You won't be able to use Grid inside React Native, though. The whole layout is based on flexbox, so you're able to use all the rules for flexbox.

You can easily use almost all the rules from CSS in one form or another. If there's something that you feel doesn't work if you're writing it in camel case, then go ahead and Google up that rule. You'll easily find out how to use almost every rule.

The `flex: 1` rule means "*let the `<View />` component take up as much space as it can*," so our **Home** screen is now the full width and height of our screen.

Let's add some new rules to our container object:

1. Add `paddingHorizontal: 32` and `paddingVertical: 64`. This will create some beautiful breathing space for us to continue styling our components.

   Let's start with our `WelcomeHeader` component.

2. We'll begin by adding `StyleSheet` to our list of imports and then create the `styles` object.

3. After that, we'll create the `upperSide`, `profilePicture`, `welcomeText`, and `levelText` styles.

4. We still can't see our picture, so let's give it a `width` and `height` of 55. To make it round, we'll give it a `borderRadius` of 55/2.

5. Now, we'll add the `profilePicture` styles to our picture via the `style` prop.

6. For our `welcomeText` and `levelText`, we'll need to specify a `fontSize` and a color, so let's go ahead and do that too. I'll use 38 for `welcomeText` and 18 for `levelText`. The color of our text is going to be set to `'#707070'`.

   We'll continue adding rules until our `WelcomeHeader` component looks like it does in our design case. Do this on your own initially. Once you've done that, check out the following code and see if you got something close to what I have here:

```
import React from 'react';
import { View, Text, Image, StyleSheet } from 'react-native';

const WelcomeHeader = () => {
    return (
        <View>
            <View style={styles.upperSide}>
                <Text style={styles.welcomeText}>Hello,{'\n'}Alin Gheorghe!</Text>
                <Image style={styles.profilePicture} source={{uri: 'https://via.placeholder.com/65'}} />
            </View>
            <Text style={styles.levelText}>Level 3</Text>
        </View>
    );
}

const styles = StyleSheet.create({
    upperSide: {
        flexDirection: 'row',
        justifyContent: 'space-between',
    },
    profilePicture: {
        width: 55,
        height: 55,
        borderRadius: 55/2,
    },
    welcomeText: {
        fontSize: 38,
        color: '#707070',
    },
    levelText: {
        fontSize: 18,
        color: '#707070'
    }
});

export default WelcomeHeader;
```

Figure 4.13 – Our fully styled WelcomeHeader component

With that, we've managed to style our `WelcomeHeader` component. I used `justifyContent` to push the image and text in opposite directions and I also specified the `flexDirection` because, by default, all the components are rendered in a column fashion. However, we needed a row for this particular example.

We won't look at styling rules any further here as you might need to discover them by yourself through practice. So, my best advice right now would be to just go ahead and get creative. Get inspiration from the apps you're using daily and create some components that look similar to whatever you picked. Try to recreate as many components as you can and see which are visually appealing to you. After a while, this will become second nature.

Don't be upset if you can't remember a certain rule or you can't think of a way to style something in a certain way. The truth is, most programmers do forget and most of them look up really basic things on Google. The most important thing for you right now is not to get upset if something doesn't work but to see that as a challenge – a challenge that will 100% improve you as a developer.

We'll stop with the styling part because we've already done it for one component, and I feel like I could show you something that might change the way you view styling. This is something that we'll start using from now on whenever we start creating apps: Galio.

The superhero, Galio

We talked about Galio at the beginning of this book. We discussed why you'd want to use it and how exactly it brings value to your app. Now, it's time to use it and see what this UI library is all about.

Right now, we need to write a different styles object for each element we're using. Galio can help solve that problem by using `props`, which will help you style your code as you develop the app.

Let's start by installing Galio to our application. For that, we need to go to our Terminal and run the following command:

```
npm i galio-framework
```

This will install the latest available version of Galio into our project. Now that we've installed Galio, let's import some components from it into our `WelcomeHeader` component.

Let's go to our `import` section and write the following:

```
import { Block, Text } from 'galio-framework';
```

If you've written this down and saved your file, then an error will appear. That's because we're importing `Text` from both `react-native` and `galio-framework`. Delete it from `react-native` and everything should work nicely again.

Oh well, nothing has changed. This is because the `Text` component from Galio is just extending your usual `Text` component. However, it comes packed with new props that will allow us to remove certain styles.

Let's delete the `style` prop on both of our `Text` elements and add `color="#707070"` instead. Now, our texts are small but they are the same color, which is cool. This means our props are working correctly. If we want to change the font size, we just have to add a prop. For our first `Text` element, we'll add h3, which stands for *Heading 3*, while for our second `Text` element, we'll add p, which stands for *paragraph.*

Now, if we hit save, we'll see how our **Text** elements suddenly have different sizes and everything looks good. We can now remove the unused style objects; that is, `welcomeText` and `levelText`.

Let's move on and see if we can remove even more. We should replace the `<View />` component that wraps our `Text` and `Image` elements with a `Block` component.

Now, let's add the following props to our newly implemented `Block` element: `row` and `space="between"`. Because of this, we can delete the `upperSide` object from our `styles` object. Now, everything looks the same but with much less code and is easier to notice.

A `Block` component is the same as a `View` component but it is packed with a lot of props that can make our development process easier.

Once we've replaced it, let's replace the other `View` element as well. We will also remove it from the imports because we don't need it anymore:

```
import React from 'react';
import { Image, StyleSheet } from 'react-native';
import { Block, Text } from 'galio-framework';

const WelcomeHeader = () => {
    return (
        <Block>
            <Block row space="between">
                <Text color="#707070" h3>Hello,{'\n'}Alin Gheorghe!</Text>
                <Image style={styles.profilePicture} source={{uri: 'https://via.placeholder.com/65'}} />
            </Block>
            <Text color="#707070" p>Level 3</Text>
        </Block>
    );
}

const styles = StyleSheet.create({
    profilePicture: {
        width: 55,
        height: 55,
        borderRadius: 55/2,
    },
});

export default WelcomeHeader;
```

Figure 4.14 – Our WelcomeHeader component with newly implemented elements from Galio

We now understand how Galio works and we'll how much it will help us moving forward with this app. So, let's go ahead and start modifying the rest of the components.

Let's go into our `MostPlayedGame` component and start importing whatever we need from Galio. Again, we need to use `Block` and `Text`. After importing these two components, we can remove the `View` and `Text` imports from `react-native` as we don't need them anymore. Don't hit save yet as we need to replace the `View` element from inside our function with the `Block` element, as we did previously. Now, you can go ahead and save your file and you'll see no changes. That's perfect – we can now start styling this component.

Let's go ahead and add the following props to our `Text` component: `size={15}` and `color="#707070"`. This will change the font size of our text and also its color.

Now, we need to import `StyleSheet` from `react-native` and use it to style `Image` so that it can be rendered on our screen. We'll create a new `styles` object from our `StyleSheet.create` method and we'll have the `image` object inside it.

After that, we'll add a `container` object as well so that we can create some space between our components. This will be used in our `Block` element.

Our new `styles` object should look something like this and have the following values:

```
const styles = StyleSheet.create({
    container: {
        marginTop: 32,
        backgroundColor: "#D8E2DC",
        paddingBottom: 4
    },
    image: {
        width: 'auto',
        height: 60,
        resizeMode: 'cover'
    }
});
```

Figure 4.15 – styles being used for our MostPlayedGame component

After writing all this and linking our `styles.container` and `styles.image` objects to the proper elements (the `Block` element and the `Image` element), we can see that our screen is starting to look more and more like the design we saw at the beginning of this chapter.

By the way, I've added 4px of `paddingBottom` to our container style just because I felt like our `Text` element could have some breathing space. We could've also created a new style for `Text` and created some padding around it. There is no *right way* of writing styles, so long as its purpose, which is to display what you want to have displayed, is respected, so have fun and experiment as much as you like.

Don't forget that we're linking our styles to each element via the `style` prop.

Oh well – I guess things are getting easier with Galio and styling as we've already been through this much, so I'll take a break and let you style the rest of the components. Once you've done that, come back to this book and see if we've taken the same path by comparing your results with mine. Maybe yours will look even better than mine while also having cleaner code and if that's the case, you should treat yourself tonight.

Have you finished? Cool – let's move on! Let's jump to our `LastPlayedGameList` component. This should be straightforward, so let's import our `Block` and `Text` components from `galio-framework` while also completely removing our imports from `react-native`. That's right – we don't need those anymore.

We'll then change the `View` element into a `Block` element. While we're here, let's add some in-line styling as well; that is, `style={{ marginTop: 32 }}`. We've added that to create more space between our components.

Now, let's go to our `Text` component and add the `color="#707070"` and `size={18}` props. And with that, we're done. We've created this component pretty quickly, right? Well, styling isn't that hard, especially when Galio is involved.

Let's move on to our last component, `PlayedGameItem`. This one will be the same thing as the previous one. We'll remove the imports from `react-native` while adding the `Block` and `Text` imports from `galio-framework`.

Now, let's replace the `View` element with our new `Block` element and add the `row`, `space="between"`, and `style={{ marginTop: 16}}` props to it. After that, we'll add our `color="#707070"` and `size={14}` props to both our `Text` elements:

```
import React from 'react';
import { Block, Text } from 'galio-framework';

const PlayedGameItem = (props) => {
    return (
        <Block row space="between" style={{ marginTop: 16}}>
            <Text color="#707070" size={14}>{props.game}</Text>
            <Text color="#707070" size={14}>{props.hours}h</Text>
        </Block>
    );
}

export default PlayedGameItem;
```

Figure 4.16 – Our fresh new component after adding Galio and styles

And with that, we've finished. Save your file and take a look at your simulator. It looks just like what we wanted. Take a moment to add more character to the screen before moving on. Change the pictures to whatever images you'd like to see there – maybe add a profile picture and an image of your favorite game.

Remember how we used props to pass down information from a **parent component** to a **child component**? You can do the same thing and change the name in our `WelcomeHeader` or even make it more modular and send all the information from the **Home** screen to your components.

Now that we've finished styling our app, let's see how we can use it on our phones.

Let's install it on our phone

We discussed why Expo is great in *Chapter 1*, *Introduction to React Native and Galio*, and I think that the people from Expo did a great job at creating that framework. The thing with smartphones is that you can't install the app very easily on your phone.

Android is a lot more open compared to iOS and you'd probably be able to export an .apk file into your phone just to have it there. However, iOS doesn't let you do that.

Of course, we could use **TestFlight**, which is an Apple service that allows you to test and share your app with other testers. But that doesn't help us because who would install TestFlight on their phone just to see your one-screen app, especially when you need an Apple Developer account?

Expo offers us a great little app called **Expo Go**. You can find it on both **App Store** and **Google Play Store**. Download it and log in or create a new account if you don't already have one. Here, you can create a build for your projects that can be tested at a later date. By doing this, we can show our friends our app without worrying too much about the other obstacles.

Publishing a project on Expo is easy; we just have to follow some steps. Let's close our development server by going into our Terminal and pressing *Ctrl + C*; then, type expo signin and press *Enter*. A message should appear, asking you for your username and password. If you still don't have an account, jump over to Expo's website and create one. After writing down your username and password, you should get the following response: **Success. You are now logged in as YOUR-USERNAME**.

Right now, there are two options available for us to use if we want to publish our app with Expo. We'll talk about both of them in the following sections as errors might happen anytime. If you encounter an error, it's best to just try an alternative method.

Publishing via Expo Developer Tools

Now that you've logged in, let's open our server again by typing `expo start` into our Terminal and hitting *Enter*.

The development server has started and a new tab containing Expo Developer Tools should have loaded in your browser. Remember that in *Chapter 1, Introducing React Native and Galio*, we showed all the available options; let's click on **Publish or republish project to the internet**:

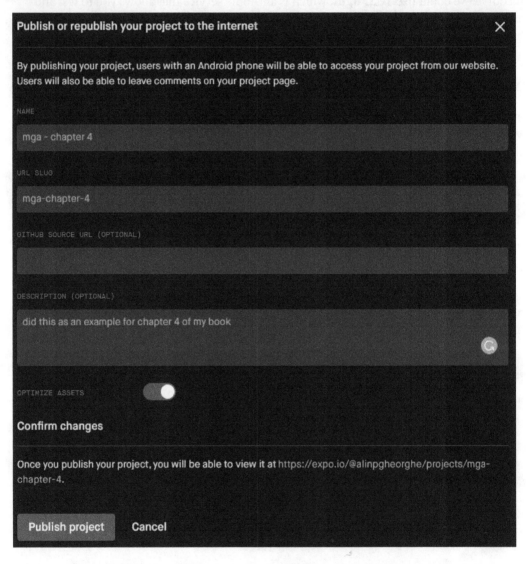

Figure 4.17 – All the information is displayed when we click the Publish button

Now, your app should be published, which means you can go inside your Expo Go app on your phone and open your app. See? Easy! Go ahead and show it off to your friends!

Publishing via the Expo CLI

Now, there may be a possibility that the first option doesn't work for you or you encountered an error. Sometimes, errors just happen and it might not even be your fault. In that case, stop our development server and write the `expo publish` command in the Terminal. A big message will appear, stating that it's going to start bundling your app and prepare it for publishing. After a while, you'll see that it has been successfully published to Expo.

Now, your app is ready to be seen by the world. Well, kind of. You could log into your Expo Go app and see your apps under the published projects category on your profile tab. The thing is... other people from the internet might see it on the Expo website and download it on their computer, but your friends won't be able to download the app on their mobile phones. That's because we haven't published the app on the official stores. It's not even on a store – it's saved in the cloud for other Expo users to see and, of course, for you to have access to it any time you want.

Congratulations! We've finally created our first complete screen. I hope you're feeling good because there's more knowledge to come that will make development even easier and a lot more fun!

Summary

In this chapter, we've experienced the process of creating a screen for our app. We took a design file, looked at it, and recreated the design with no functionality. This is a great step in anyone's career as this is your first time finishing an app idea. I think you should pat yourself on the back and realize that what you're doing here is not that easy. A lot of people would not even try to start learning about this, but you've done it. On top of that, you've even created a fully styled screen.

Once we'd learned about styling, Galio came in. We learned how building a layout with Galio makes it a lot easier for us. We're still not escaping the styling part entirely but we would never be able to not style something. After all, styling is fun. By using Galio, we've seen how easy it is to arrange elements and create fast prototypes.

By the end of this chapter, we looked at two different ways of publishing our app idea to Expo Go, a mobile app that helps us play with our project without actually pushing it to the store. That's cool and I bet your friends and family will be overjoyed with seeing how much progress you're making.

Now, it's time for us to move on to the next chapter where we will discuss the benefits of using Galio.

5
Why Galio?

In the previous chapter, we created our first screen. After creating it with plain React Native code, we went ahead and imported some Galio components, which helped us style and create the layout in a much easier and simpler manner.

This chapter is going to be a better introduction to Galio. We'll learn how to use it and why most programmers look for a **user interface** (UI) solution such as Galio to solve most of their developing processes. As we saw in the last chapter, just using the core `react-native` components means the code gets really large and hard to maintain. Galio components come packed with many different props that make life a lot easier.

We'll also learn about the benefits of using an open source library, how that creates a community of people willing to help each other, and how you can step in and add value to the library as you see fit.

This conversation is going to open a lot of new doors that you never thought even existed before. It will create some sort of new mentality, expanding your vision on what a developer really is and how they communicate.

We will cover the following topics in this chapter:

- Beautiful mobile app development with Galio
- Using Galio in your apps
- Discovering the benefits of Galio

By the end of this chapter, you should be able to understand the reasons why people choose Galio to quickly start working on their projects. You'll understand how to install it and use it in your apps and what role certain components have in your app. Learning about certain components sure is helpful, but don't shy away—all programmers use Google to discover solutions to their problems, and I highly encourage you to do the same if you feel some things require further explanation or have changed during the course of time.

Technical requirements

You can check out this chapter's code by going to GitHub at `https://github.com/PacktPublishing/Lightning-Fast-Mobile-App-Development-with-Galio`. You'll find a folder called `Chapter 05` that contains all the code we've written inside this chapter. In order to use that project, please follow the instructions found in the `README.md` file.

Beautiful mobile app development with Galio

We have so many examples of mobile applications that don't look really good. Multiple different social media apps are being created by random people thinking that they'll hit the next jackpot, just like Facebook did. The most usual problem I've identified with most of these apps, on top of the bugs you're always going to find inside a beginner developer's app, is the design.

The layout is quickly created and they do not pay any attention to how the **user experience** (**UX**) might end up because of their design. They think that just because they have a nice idea, they don't really need to pay attention to anything else.

I disagree. I honestly believe that you could sell anything as long as the design is sleek and the UX is top-notch. The reason why I believe this is mostly that I usually use the *1-minute rule*. This is something personal that I've created for myself. Basically, once I install an app, I only take somewhere around a minute to try to see what's going on with that specific app.

Why a minute? Well, we're using mobile apps because we want things to be fast and easy to use. We're always looking for a mobile app alternative to our web-related activities just because we want to have easier and faster access. We want to be able to check some information and maybe do some activities. If I can't figure out how to use your application in 1 minute, then I'll uninstall it.

What does that say about the way we should build our apps? We should have the user in mind at all times and only use enough information for them to not have to think about how to use our app.

Galio comes in handy because it uses the same process—easy, fast, and straightforward. You don't have to think too much about how wide a button should be or what the pixel size should be in your app. It comes pre-packed with all the tools you need to design and produce your best idea.

Let's start by seeing how buttons look and the many different ways of styling we have for them. We'll be using buttons in almost any situation inside our app, so I figured that this would be a great start. Have a look at the following screenshot:

Figure 5.1 – Buttons displayed inside an app

As you can see, we have lots of different ways of displaying a button, from bright colors and shadows to shadowless and plain. You could have squares, or even—straight up—just a circle and an icon.

Let's take a look at how easily this can be implemented in our app. Here's the code for the buttons shown in *Figure 5.1*:

```
<Button capitalize size="small">small size capitalize</Button>
<Button round uppercase color="error">round uppercase</Button>
<Button onlyIcon icon="tags" iconFamily="antdesign" iconSize={30} color="warning" iconColor="#fff" style={{ width: 40, height: 40 }}>warning</Button>
<Button color="#50C7C7" shadowless>custom color and shadowless</Button>
<Button round size="small" color="success">round and small</Button>
```

Figure 5.2 – Code for the buttons in Figure 5.1

So, as far as we can see, we have props for everything. Some require more or less work from us, but at the same time, the ease of just editing everything from inside the component is worth it every time. Just because we wanted to capitalize a button and have it capitalized all the time, we can use the `capitalize` prop. Or maybe we want the text to always be uppercase; that's fine—we've got a prop for that as well. It really makes the developing process incredibly easy and accessible to anybody.

We've discussed how an app should look and feel good for our users. But that should be transposed into our development process as well. That's why I honestly believe that clear and beautiful code will almost always equal a beautiful product.

Now, let's take a look at another cool component from the Galio package—the `Accordion`. You never know when you need a beautiful-looking accordion to create more space for your content. You can see a representation of this here:

Figure 5.3 – Accordion component as displayed on your screen

This component is incredibly easy to use. It needs a `View` (or `Block`) component with a specified height and an array of objects defining the content inside the component. Here's the code relating to this component:

```
<Accordion dataArray={data} />
```

This should basically be pretty easy for anyone to configure and use when needed. The objects inside the array must have certain keys so that our component can identify and understand where to place that content. An object might look something like this:

```
{title: "2nd Chapter", content: "Lorem ipsum dolor sit amet"}
```

It doesn't look that hard, right? If we want to use an icon, as we saw when we created our first item (in *Chapter 1, Introduction to React Native and Galio,*), all we have to do is add the icon key inside our object, which is going to contain the name, family, and size values of the specified icon we want to use.

This is basically how all of Galio's components are built. They are straightforward and good-looking, ready to be used to create a new app in seconds.

We should check out another component before moving forward, so let's see how easy it is to customize and use a checkbox with Galio. Have a look at the following screenshot:

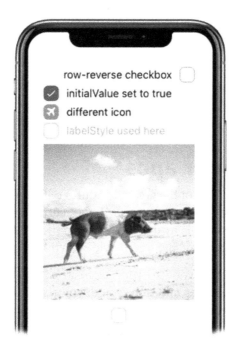

Figure 5.4 – Checkbox component as displayed on your screen

This looks complicated if it is to be created with the core React Native components, but we're in luck because Galio makes it as easy as just writing a single line, as illustrated here:

```
<Checkbox color="primary" flexDirection="row-reverse"
label="row-reverse checkbox" />
<Checkbox color="info" initialValue={true} label="initialValue
set to true" />
<Checkbox color="error" initialValue={true} label="different
icon" iconFamily="font-awesome" iconName="plane" />
<Checkbox color="warning" labelStyle={{ color: '#FF9C09' }}
label="labelStyle used here" />
<Checkbox color="success" image="https://images.unsplash.
com/photo-1569780655478-ecffea4c165c?ixlib=rb-1.2.1"
flexDirection="column-reverse"/>
```

As we can see, we can set the color, the direction, and even the icon, all inside our component. They look beautiful to both write and display as they are without any modification. This makes us proud of our library as we really take pride in the way it looks.

Now, let's take a look at how easy it is to create a basic layout with the Block component. We'll be using this component for layout design only, and we'll color every square so that we have a better understanding of what each element is exactly displaying.

Using Galio in your apps

Now, let's see Galio in action. One of the greatest features of Galio—besides the way it looks or the ease of writing code—is our Block component, which is basically a View component but with superpowers. Why do we say superpowers? We could easily use this component to both create our layout and easily style everything just by using props.

So, let's put this into action and see how easy it is to create a basic layout with the Block component. I'll take you step by step and show you the most usual ways of using Block while also demonstrating the most common ways of arranging the layout. You can find the project in our GitHub repository or you can follow along by coding with me.

I'll start by creating a new Expo project. After that, I'll install Galio via the command line by writing the following code:

```
npm i galio-framework
```

Now that we have everything installed, I'll skip the whole process of organizing our files as we're using this for demonstration purposes. So, we're going to be writing our code directly into our `App.js` file, in our entry point—the `App` function.

We're going to import our `Block` component via the `import` function under our other imports, like this:

```
import { Block } from 'galio-framework';
```

I'll delete everything inside the `App` function and I'll start by creating my first `Block` component. This will be used to keep all the elements inside because, as we know, we can't return more than one component in a function, so in conclusion, that one component will have to encapsulate the other components inside it.

We'll use the `flex` prop on it, which will make our `Block` have the property of `flex: 1` so that it will stretch both horizontally and vertically, covering the entire screen.

Now we're done with this, let's use the `style` prop. As we said, each `Block` element will have a `backgroundColor` property so that we can more easily identify which one is which. Inside our `style` prop, we'll write `styles.container`.

Remember that we have a `styles` object below it that has all the styles we can use via the `StyleSheet.create` function. We'll delete everything inside the container there, and we're only going to write `backgroundColor: '#F94144'`.

Let's save, and now, our screen should be some sort of red. Fun fact, this color is called **Red Salsa**.

Now that everything is working, let's go ahead and start creating our layout of boxes and see how easy it is to arrange elements inside our app with the `Block` component.

By the way, you should also remove unnecessary imports such as `StatusBar`, `Text`, and `View`.

We'll now start by creating three `Blocks` inside our main `Block` component. As we know, all the components inside React Native are arranged in a column from top to bottom, so we're basically going to create three rows of `Blocks`.

Each of these rows will have the `style` prop on them, and in order from top to bottom, the styles will be called `styles.row1`, `styles.row2`, and `styles.row3`. Now, we'll go inside our `styles` object and create row1, row2, and row3 styles. Each of them will have only one property and that is `backgroundColor`, with values in order from row1 to row3, like this: #F3722C, #90BE6D, #277DA1.

Now, if we save, we'll see nothing. Well, that's because our `Block` element does not have a size set, so it doesn't know how much space it needs to occupy. Remember what we did with the last one? We used `flex`, so let's use the `flex` prop and all three of our components, as follows:

```
export default function App() {
  return (
    <Block flex style={styles.container}>
      <Block flex style={styles.row1}></Block>
      <Block flex style={styles.row2}></Block>
      <Block flex style={styles.row3}></Block>
    </Block>
  );
}

const styles = StyleSheet.create({
  container: {
    backgroundColor: '#F94144'
  },
  row1: {
    backgroundColor: '#F3722C'
  },
  row2: {
    backgroundColor: '#90BE6D'
  },
  row3: {
    backgroundColor: '#277DA1'
  }
});
```

Figure 5.5 – The code we've used to create our three rows

Hit **Save**, and we suddenly see three colors from top to bottom: orange, green, and blue; more exactly: Orange Red, Pistachio, and CG Blue.

Because of the `flex: 1` property that gets applied when we use the `flex` prop, each of them gets equal space inside the main `Block` component. Now, the cool thing regarding this `flex` property is that we can use it to set the amount of space we need.

Let's go ahead, and for the first row, we'll set it to `flex={2}`; for the second one, we'll leave it as it is; and for the third one, we'll set it to `flex={3}`. Now, we can see that each box has a different amount of space allocated. This is all thanks to the fact that React Native uses a **flex system** to create the layout; we're just profiting from how easily accessible it is to use it with Galio.

Now, let's see how it does the math when we set all these numbers to the `flex` property. Because we've left the second one as it was, that will be transformed when rendered to `flex={1}`. We'll do the math between the three flexes and end up with the following: `2+1+3` = 5. So, in a nutshell, we can say *the first row is two parts of five*, *the second one is one part*, and *the third one is three parts*. The numbers used here are specific to our app, but you might have different numbers. The main idea is to understand the fact that those numbers are dividing the space they have at their disposal—a bigger number gives us more space while a smaller number gives us less.

Now, let's use the first row to put another set of `Block` components and use more props. Yes—we do have a lot of props to go along with this.

We'll start by typing out one component for now and create a style called `row1el`. Apply that style to our new `Block` and use the `#577590` color. Well, yeah—nothing shows up, but let's use two more props to make it show up. We'll write `width={50}` and `height={50}`. This will set the width and height of our `Block` component in pixels.

Let's center this element by using the `middle` prop on the parent component. The parent component is our first row. As you can see now, our dark blue `block` element is in the middle of our first row:

```
export default function App() {
  return (
    <Block flex style={styles.container}>
      <Block flex={2} middle style={styles.row1}>
        <Block width={40} height={40} style={styles.row1el}/>
      </Block>
      <Block flex style={styles.row2}></Block>
      <Block flex={3} style={styles.row3}></Block>
    </Block>
  );
}
```

Figure 5.6 – Our code with the newest elements inside of it

Now, for the second row, let's go inside our `styles.row2` object and add padding. We'll add `padding: 30`, and we can observe how our second row got suddenly taller. That's because our whole layout (the three rows) is built with flex, which is not setting an absolute size in pixels; the component now wants more space.

Inside our second row, we'll create another `Block` with props of `flex`, `middle`, and `style={styles.row2gal}`. Now, for our `row2gal`, we'll have `backgroundColor:` `'#F9844A'`. Let's add three `Block` components inside of this one. Each of them will have the following props: `width={30}`, `height={30}`, and `style`. The styles will be named in order, from top to bottom, `row2p1`, `row2p2`, and `row2p3`. Following the exact order of our styles, we'll have for each of them the `backgroundColor` property set to `'#4D908E'`, `'#43AA8B'`, and `'#F94144'`.

Now, if we hit **Save**, we'll see that our `Blocks` are positioned in a column. Let's solve this by using in the parent component the `row` prop. Now, we've got them in a row—that's pretty cool, right? Let's use the `middle` prop as well, and `space="evenly"`. Save and see how it looks. Our elements are now centered and they have even space between them and the left and right margins of the parent component.

Now, let's go to the second `Block` and use the `bottom` prop. This will make the second element go below the first and third ones. Kind of funny—it looks like a face, right? See if you agree:

```
export default function App() {
  return (
    <Block flex style={styles.container}>
      <Block flex={2} middle style={styles.row1}>
        <Block width={40} height={40} style={styles.row1el}/>
      </Block>
      <Block flex style={styles.row2}>
        <Block flex middle row space="evenly" style={styles.row2gal}>
          <Block width={40} height={40} style={styles.row2p1} />
          <Block width={40} height={40} bottom style={styles.row2p2}/>
          <Block width={40} height={40} style={styles.row2p3}/>
        </Block>
      </Block>
      <Block flex={3} style={styles.row3}></Block>
    </Block>
  );
}
```

Figure 5.7 – Our code after we've filled the second row

You can see how easy it is to create a basic layout by only using `Block`. Now, before moving forward, you should take your time, and instead of `bottom` maybe use the `top` prop on another component and see how it works. Or, instead of `space="evenly"`, you could use `space="between"` or `space="around"`.

This becomes fun really fast as we actually have full creative control by using those components. The best part of this is you can create a full screen made out of `Blocks` and then just populate every `Block` element with the component that you want to have. Honestly, these features alone would be enough for me to start loving Galio. Good thing we have even more features.

Now that we've used some Galio features in our app, let's move forward and look at what type of benefits Galio offers.

Discovering the benefits of Galio

Now that we've been through several of the benefits of using Galio—such as the ease of writing code, how beautiful it looks, and how cool it is to create a layout with it—we're ready to see other benefits of using it, and I feel the best place we should start our journey is GitHub.

You can see the Galio icon here:

Figure 5.8 – Screenshot taken from Galio's landing page

As I've said, we're lucky to have this great community as there's always someone reaching out to help you. You can also help out other people, which we always encourage. I feel that Galio's community might be best defined by the word *collective*. In the music industry, this word is mostly used to define a group of people with similar interests that just work together and help each other because they know that more people means faster and easier development for everybody.

Let's take a look at some ways you could help and be part of this community.

First of all, we have the Discord server, which is where most of our developers hang out and discuss random things but also bugs and how to solve specific questions. This place is basically a big chat room where everybody is having fun.

Anybody can join it and ask questions, or even report a bug or something that is not working. Maybe you feel the design could be improved and you want to pitch a whole new look to Galio and its community. You can do that there with no worries of someone laughing at you or not taking you seriously.

On top of the Discord server, we have the GitHub repository and the website. The GitHub repository is where we keep everything that's code-related. This is where we maintain the code, answer issues, create new development plans for the future, create hotfixes for certain products, and work with **pull requests (PRs)**.

A PR refers to when someone wants to help out with a library. So, they start by creating a **fork**, which is the act of cloning someone's repository. Then, they make their own modifications, and then the new copy of the repository is submitted as a **PR**. that will be then verified by an admin and accepted or rejected depending on if the code respects the rules and if it's part of the development plan.

Our website is mostly where we want to showcase people's apps and news about Galio. It's where we present Galio to the world but it's also where we keep a really important part of the whole library: the documentation.

The documentation is your go-to place anytime you want to learn more information about a specific component or how to use a feature of Galio, such as, for example… the **GalioTheme** feature.

Everything related to Galio—such as the colors, sizes, and layout rules—is stored in our default theme. This can be found in the `theme` folder inside our library. Every component inherits its styling rules from that file. The coolest thing is that you can actually rewrite our theme file with only the things you want to modify by using our theme components.

For example, let's say you want a different color code for `primary`. You can override our primary color with your own color and use it with Galio as though it's always been there.

To use the GalioTheme feature, you'd have to import `theme`, `withGalio`, and `GalioProvider` from our library. Let's take a small example here:

```
const customTheme = {
  SIZES: { BASE: 18, }
  // this will overwrite the Galio SIZES BASE value 16
  COLORS: { PRIMARY: 'red', }
  // this will overwrite the Galio COLORS PRIMARY color #B23AFC
};
  <GalioProvider theme={customTheme}>
  <YourComponent />
</GalioProvider>
```

This will create a `customTheme` object that will contain two keys: `SIZES` and `COLORS`. If you only want to modify the colors, you can use just that specific key. Then, you need to encapsulate your component with our **higher-order component** (**HoC**), called `GalioProvider`. We'll also need to pass our new `customTheme` object to Galio via the `theme` prop.

> **Tip**
> An HoC is an advanced React feature that can be more easily defined as a function that returns a component and improves that component in some way. Let's say you're Tony Stark, and the HoC is the Iron Man suit. The suit is made of iron gloves, boots, armor, and helmet, and Tony with iron boots can fly.

Now, the `customTheme` constants will overwrite the default Galio theme constants.

But wait—maybe you don't want to change our theme but you want to use our constants inside your styling. Using our design system might help you design your layout faster, and we're always using Galio's constants inside different products we're creating for our clients.

Exporting a React component using the `withGalio` function enables your component to consume Galio's React Context and pass down a theme in your component as a prop or as an argument for the `styles` object. Let's take a look at how this is done—I'm sure you'll understand it:

```
const styles = theme => StyleSheet.create({
  container: {
    flex: 1,
    backgroundColor: theme.COLORS.FACEBOOK
  }
});
export default withGalio(App, styles);
```

Because we're using the `withGalio` function to export our component, Galio will pass down to the object we've selected (in this case, it's `styles`) all the constant theme variables we have inside our library. So, that's why we're able to use `theme` as an argument inside our `styles` object and change the `backgroundColor` property to that Facebook color we have inside our library.

You'll find a table with all the information regarding our constants on our documentation website, which is at `https://galio.io/docs`.

As you can see, Galio is fully packed with lots of cool features that will help us develop any mobile app extremely fast and, in the end, make it really good-looking. So, why not give it a try? We'll code all of our projects with Galio from now on. This will be a mandatory import at the beginning of each app we'll do from now on in this book. We'll use more of Galio's components than React Native's ones.

Given this, we'll learn more and more about how to use Galio and how to design great apps until we can start coding our own ideas. Maybe one of us will actually create a great app with a lot of value for society—something that will change the world.

It's nice to dream about how many things we'll be able to do once we acquire more and more knowledge. This daydreaming and constant focus on your objective is going to prove one of the greatest weapons in learning how to code.

Summary

In this chapter, we've been through multiple examples of why Galio is such a great library. By the end of it, you must've figured out that Galio really deserves to be one of the libraries under your belt—one library to rule them all. This will act like your main package with which you'll create incredibly stunning apps, both visually for our users and our programming buddies who want to help us out with the code.

Don't be afraid of looking into Galio's core code. You might learn a lot of things from just experiencing and understanding Galio's code. You might even be able to create your own library.

So, we've discovered that Galio is really cool because the code is easy to use. We just have a few props that can change the whole world in terms of speed of coding and easy access to specific parameters. We've also seen how great Galio looks out of the box. I mean... this library is gorgeous. Sometimes, I wouldn't even edit the styling; I'd just use Galio styles because of how great they look.

We've also seen how easy it is to create a layout with the `Block` component and how placing objects on the screen is a lot easier than we thought as long as we know just a few props that go along with the `Block` component.

After that, we discussed what a great community Galio has and how we can take part in it. We haven't gone too deep into GitHub as this is out of the scope of this book, but we've definitely learned a lot about how that community works and how we can take part in it.

At the end of this, we discussed some more advanced features of Galio—or, to be more correct, features that use more advanced features of React because they're really easy to use if we want to use them from Galio.

In the end, we can say that Galio creates an easy access route for everybody into the mobile developing world, and I think it's safe to say we're all grateful for its existence.

The next chapter is going to cover the basics of the mobile UI. We'll figure out how to build a clean-looking UI for our apps while learning some guidelines and rules on how to provide our users with the best UX we can create.

6
The Basics of Mobile UI Building

Now that we understand more about how Galio can help us with building our cross-platform mobile application, it's time to learn a few rules about designing so that we can use the framework to its maximum potential.

This chapter will look superficially at some design concepts and guidelines that'll help us at least feel more confident in our design skills. I hope this chapter will give you a boost of confidence and a drive to build/create a beautiful **user interface** (**UI**).

We'll start by exploring the importance of clean design and some basic guidelines we should follow to make sure our design is as clean and minimalistic as possible while delivering the most useful information to our users. After that, we'll slowly drift into a basic explanation of **user experience** (**UX**) and how to find out what's best for our users.

Once we've discussed all this, we'll find out how we can minimize user input so that our users won't feel like dropping a form in the middle of completing it. We'll go over what exactly might block our users from finishing a form and how we can improve our forms so that the completion rate might increase.

After that, we'll see how decluttering our design ideas is usually the best way to make sure our app will look clean and good. We'll discover guidelines on how to do that and what are the most appropriate ways of creating breathing room. We'll also go over my creative process, from the first design draft to a final screen that I feel deserves to be implemented.

After decluttering, it's time to talk about consistency. We'll learn how and why consistency is important in a mobile application. We'll also go over the main ideas of three different UI design tools so that you'll be able to at least know what to research and form your own opinion on which to pick to prototype your own applications.

The following topics will be covered in this chapter:

- Exploring the importance of clean design
- Minimizing user input
- Decluttering for better app organization
- Maintaining consistency in your app

Exploring the importance of clean design

Now that we have got to this step, it's time for us to learn a few rules and guidelines regarding how to create a good-looking design for our app. Now, *beauty is subjective*, as we all know, but there are certain rules that may create a better flow inside your mobile app or even your website.

We are not trying to look through objective lenses toward what beauty is, but there are certain aspects of beauty that are in direct correlation with our brain and how it's built. For example, *colors can mean different things in different cultures*, and that's OK as we're not going to get into choosing yellow over black.

At the same time, we can take the *rule of thirds*, which is a rule of thumb for creating visual art such as films, paintings, or photographs. The thing about the rule of thirds is that we've discovered that for some reason, our eyes pay more attention to the subject of a photograph when it is sitting at the intersection of two lines after we've divided the screen into two-thirds, both horizontally and vertically. You can try this out yourself right now. The chances are your phone has this feature already built inside your camera app, so try to take a picture with the subject at the center and then with the subject where the lines intersect.

This, of course, does not mean that all of our pictures must be taken following the rule of thirds but it will help in most cases. The thing is that there are other factors that would have to be taken into consideration, such as shading, contrast, brightness, and so on.

The whole purpose of explaining this is for you to understand that there are certain aspects of a design that can manifest the idea of "beauty" to our users.

One of these is actually the importance of a clean design and how it generally helps us convey our mobile app's purpose in a straightforward manner. I'm not trying to create a minimalism-loving group, but I do feel that in this day and age, minimalism has become more important. In a world where there are so many choices to be made, the user loves it when the information is straightforward and they don't have to scroll through a big screen of information and distractions to get to the point of your website.

"Design is not just what it looks like and feels like. Design is how it works"

– Steve Jobs

So, what exactly does it mean to have a clean design? Remember those websites back in the 2000s, or even 2010s, that were full of unnecessary information, such as a clock randomly placed in the right corner of the screen? People don't want to see your app or website crowded with… stuff. They actually prefer a more simplistic approach that makes it look more stylish and cooler and avoids sending the user through multiple pages of nonsense just to get to their objective, which might be just finding out where your company is located.

Let's implement some rules so that we won't have to deal with that sort of nightmare application.

Essentials

Focusing on the essentials allows us to keep things short and to the point. We can do that by limiting the number of visual elements and menus. If you think about a drop-down menu inside your mobile app, you'd better stop thinking about it right now and maybe start thinking about how you can divide your app into categories that work with something such as a bottom tab navigator.

Color scheme

Let's be honest. We all love colors! We do! They're pretty, and whenever we go out with our friends clubbing or just having fun in general, we're always stressing about what should we wear. Well, that's because not all colors work together, and sometimes if you pick more than 10 colors, people won't have an idea what to focus their eyes on.

The same thing goes for websites and mobile apps. We should limit our color usage to only three colors—of course, while applying different shades where necessary—but by having only three main colors inside our app, we could create some sort of continuity inside our app.

Let's say one screen has a **Submit** button that is green. The other one has a **Submit** button that is purple. Once the user sees that, they'll instantly think: "Is that the correct button?" Once you create a rule for how certain things should look, stick with it!

Availability and accessibility

This one is actually really critical. Your mobile app's design has to be able to function on all distributing platforms that are of interest to your target audience. I'd actually say that right now, in today's market, it is mandatory to have your product on at least iOS and Android.

The thing is, because these platforms were built differently and have a different UX, you'd have to adapt your product for each platform. The better you do this, the more people are going to enjoy using your app.

Also, we're in 2021, so you should implement support for screen readers such as **VoiceOver** for iOS or **TalkBack** for Android. This will make you feel better about yourself because you're not only creating a better digital world for everybody but also, by allowing a bigger audience for your app, you'll have a better chance of developing your idea into a successful one.

Simplicity

I can't stress this enough, but you need to focus on what's important. You don't need to list every single thing your app can do in a single screen. Try to keep it short. No one has the time to actually read all the information on a page, so having as little information as possible but at the same time making this as meaningful as possible is the key to simplicity.

Information architecture

Every user that's going to interact with your app has a pre-built behavior pattern that's going to be exhibited when they first use your app. Study your competition and make sure that behavior is not going to get in the way of your creative process while making the app. For example, they might expect a specific button, such as the **Get Started** button inside a splash screen, to be always at the bottom of the screen. Your job is to make sure you'll use these behaviors in your favor, and if you want to create some sort of new UX for your users, take your time to teach them how to use your app.

Consistency

Make sure your design and information are consistent throughout the whole app. By staying consistent, you can make sure your users will never have a moment where they do not understand what's going on or how to use your app. By staying consistent, we actually teach our users the best way to use our platform without the need for extra boring text.

User experience

You've probably noticed the phrase **user experience (UX)** , but we haven't really defined it. UX refers to how well a product (website or mobile application) meets the needs of a user.

We should *distinguish UX and usability* as the latter is a quality attribute of the UI, covering how easy the system is to learn or how efficient it is to use.

A good rule of thumb we can keep in mind when designing a mobile UX design is to ask ourselves the following question: *Is the mobile app useful?*

If not, we can say there's no value for the end user.

If the answer is yes but it's not intuitive enough, the end user won't spend time learning it.

Mobile UX design encapsulates three important aspects: accessibility, discoverability, and efficiency. This is leading to fast, positive, and experience-driven end results.

Net Solutions' State of B2B Commerce 2020 report states *65.8% businesses will be investing in improving mobile UX design in the next 12 months.*

Based on this data, we should realize that UX is a never-ending science. We'll never have the perfect UX—this will change as the user changes. In time, we'll likely change our behavior while using our phones, so your designer needs to design a great experience for your end users that will meet the expectations of those specific users at that specific time.

The natural question right now should be: *How we should approach this so that we'll always be able to deliver a high-quality user experience to our users?* I'd say for constructing a great UX, the approach mentioned in the following sections will get you the best results.

Research

Spend multiple days with your end users. Understand their needs and how exactly they feel about the way things are working right now. Hear them out because their feedback is one of the most important parts of this whole process.

For example, if you take a look at one of the older members of your family when they're using an app, you'll notice they get frustrated really easily with certain parts of the application. Watch them and check for what they're expecting. They might say something such as "Why is it so hard to order something through this app?" and proceed to randomly tap on the screen nervously because things are not exactly how they're expecting them to be. People around you and especially those that are targeted by the app can offer you the most valuable pieces of information.

Empathize

After discussions with the users and understanding their needs, it's time to find solutions to their problems. Use whatever helps you the most to organize those thoughts and try finding solutions as to how to eliminate problems your users have encountered. You need to pay attention so that you don't create more problems, so test your app after you find a solution.

Build

Well, this is self-explanatory. Once everything has been tested and you have found solutions to all the problems, it's time to build the app. The thing is… from your research, you should be able to realize what types of technologies are truly needed for your app. Sometimes, even React Native just won't cut it, so you might have to make some changes. This is a part of being a great programmer, so don't worry! Once you get to know a programming language and a framework, you can learn anything.

Now that we've been through this and we've understood a little about why a clean design is actually really important for our app and how UX works, we should have a pretty good idea as to why certain apps take the minimalistic route and have a straight learning path for all their users.

Now, let's discuss why is it important to minimize user input, which is another part of making sure we have a clean design, and how we can do it.

Minimizing user input

Lots of people are hesitant to fill out forms, especially when they're long, filled with personal information involving them having to search through physical documents, and filled with required steps that seem irrelevant.

Knowing this, we have an obligation to create a good form for our users so that they won't feel like it's a chore when completing it. The primary goal of any form is **completion**. For that, we'll first have to know what the primary concepts of an effective form are. These are covered here:

- **Perception of complexity**: Every time we're presented with a form, the first thing we're doing is visually scanning it so that we can estimate how much time is required to get to the finish line. Knowing this, we can pretty much instantly realize that perception of complexity plays a crucial role in finishing a form. The more complex it looks, the less likely it is that users are actually going to complete it.

- **Interaction cost**: This is the collective sum of all the effort put into completing a form. The more effort the user puts into it, the less likely it is they'll complete it. Imagine a form with a bug where you can't add your birthdate or it's not intuitive to do so. You'll probably lose focus and get mad at the form and how difficult it is to use. In the end, you'll never actually complete it. This type of faulty interaction makes the user think less of the app and the form itself. This is the type of flaw that will make the user forget about how beautiful the design is or how useful the rest of the app is.

Now that we know how users are actually going to think about our forms, let's see which guidelines we should follow so that we'll create an efficient form design, something that all of our users will be able to follow and complete. Consider the following points:

- **Remove user effort by asking the right questions**: Questions inside the form should be kept in an intuitive sequence and they should seem logically sorted from the user's point of view. When thinking about the sequence of asking questions, we always start with name, birthplace, and personal information. That's because this functions just like a conversation. Don't compromise only because your database or application logic has a different order of asking questions—the user comes first. Our job as programmers shines best when the user has no understanding of how the app actually works.

  A good rule of thumb for this might be to continuously ask yourself why and how the information you've requested is being used.

- **Single-column layout**: The biggest problem with a two-column layout for our forms is that you don't really know how the user is going to read the information. To make this easier, having one single column should be intuitive enough for the user to understand that they first have to complete whatever questions are at the top of the screen.

- **Use as few input fields as possible**: Imagine this—you want to book a flight and it's asking you for information regarding everything that's going to happen in your journey. You just want to check the prices and see if you can afford next month's flight to the Bahamas, but you're seeing a form just as big as your entire screen. You'll look at the screen and think that maybe you don't really want to go to the Bahamas, at least not with this booking company.

  Using as few input fields as possible not only means removing unnecessary questions from your form—you should also think of different ways of asking those questions. Instead of having three input fields for the departure date (day, month, year), it might be easier to have a date picker with a single input field. Another good example of using other types of form elements might be instead of having a dropdown for the numbers of passengers, we might use just a + and − button. This will make the content more interactive and less threatening to a user trying to quickly get through the form.

- **The correct width of an input**: This happens way too often. I'm placing an online order for something and then they're asking me for a street address and street number. That obviously means that I should write the street name in an input field and then the street number in another field. The problem is that the street field is extremely big. This makes me confused, asking myself: "Should I write anything else besides the street name?" This shouldn't happen; if you know the user is supposed to write a ZIP code, try to make the ZIP code input field just as big as it needs to be. Making it larger than it has to be might make the user confused, and we don't want to confuse our users.

- **Labels on the top**: Having the input labels on top of the text inputs makes it easier to follow the form. Let's say we have them on the left side of the screen; this is going to make your eyes go zig-zag, which doesn't seem like too much work, but we're trying to have the cleanest and most straightforward design possible, so everything that can help our users feel like our form won't be too difficult is going to be in our favor.

- **Optional and required fields**: As we know, we should always try to avoid optional fields in forms as they make the form longer than it has to be, but there are some cases where some optional fields are needed if we're trying to obtain more information for our marketing team, or maybe we just need a second address for a checkout form. If they're necessary for us to implement, then we might as well make it really obvious that they're optional and not required. You can do that just by writing **Optional** next to the label, but make sure it's visible and in no way a hidden message.

- **Highly visible error messages**: I actually hate it (and I'm not the only one) when I get something wrong in a form but I have no idea what it is. Everything becomes a puzzle: "Is it the password?" "Is it the email?" "What did I get wrong?" Avoid this by having clear visible error messages for each input form.

  The messages must be visible just by scanning the screen with your eyes. For this, you can use anything at your disposal, be it icons, colors, or text.

  The right time to inform a user about something being wrong is *after* they're done with the form. Don't interrupt their form completion process telling them they got something wrong as that might be quite annoying for some users.

Using these guidelines should ensure that we have a really good form. But this doesn't stop here. Each situation is different, so don't be afraid to break rules or ideas. The cool thing about designing an app is that your ideas matter just as much as anyone's. The best thing to do when you're trying to be different is to always ask yourself: "How does that improve my user's experience?" If you can't find an answer, it's better to just stick with these main ideas or find new ones from design books or psychology books.

So, up to now, we've discussed having a clean design and a good-looking form for our users. We should start thinking about another aspect of creating our design—decluttering.

Decluttering for better app organization

There's always an issue between displaying relevant information to the user and keeping the UI as clean and as minimal as possible. When we say decluttering, we're referring to the visual and readability aspects of design.

Clutter is terrible on desktop websites, but it's even worse on mobile apps as the screen size is a lot smaller. It's essential to get rid of any information that's not absolutely necessary.

So, let's see how we can do that with our apps. We can refer to our first screen created back in *Chapter 4, Your First Cross-Platform App*, reproduced here:

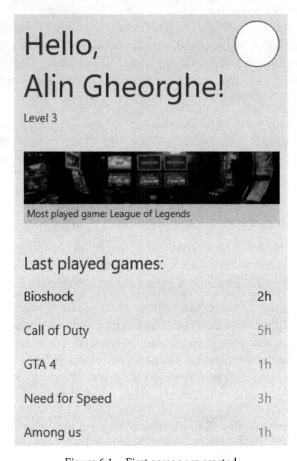

Figure 6.1 – First screen we created

As you can see, our app is already filled with only one important piece of information for our user: the last played games, the most played game, their name, and their level. But when I first started developing the idea I had for this screen, I actually started with a screen full of information. My screen looked really cluttered with information that wasn't necessary, but for some reason, I thought that might be relevant to our user.

Let's see how it looked before and try to notice how exactly I got it decluttered into the final form, as follows:

Figure 6.2 – Our screen before decluttering

I know—your first reaction is "yuck", and that's completely understandable. This screen looks filled with too much stuff. On top of that, it feels like there's no breathing room, and the information takes all the space available.

Let's take it step by step and see what I've done from my initial idea (*Figure 6.2*) to my final product for us to code (*Figure 6.1*). We'll try to understand what exactly happened inside my creative process and how we decluttered our screen. Here's how it evolved:

1. **Whitespace**

   The space between the edge of the screen and the main content area is called *a margin* in typography. Even if you write a Word document, there's always an empty space; we're not writing things from one edge of the paper to the other edge. Knowing this, even though I had a margin of 8 **pixels** (px), it still didn't feel right. I felt that more space was needed, so I increased the margin to 32px.

   This constrained our content and left us with less space to work with, but everything then looked like it got more breathing room. This is a fair exchange; less information is not always bad, especially after assessing the content of your screen.

2. **Removing unnecessary information**

   Once we identified which pieces of information were not absolutely needed for our user on this specific screen, it was time to remove them. At the beginning, I thought a nice little chart would be cool, but seeing how much space it took made me realize that for my hypothetical app it would be better to think of that chart as something that the user could see by tapping on the game they were interested in.

   The same thing applies to days they last played those games on. Those aren't needed at first glance as they could see them on another screen once they're actually interested in statistics for that game. All this information could be easily implemented on another screen, so why would we place it on the first screen?

3. **Alignment**

   Now that we've removed some elements and we've chosen the 32px margin from the edge of the screen to our content area, it was time to create some rules for alignment. First of all, I thought to myself that we should have everything aligned to the left side of the screen. If we were to break that rule and suddenly have a heading text on the center of the screen, our users might think there's something off.

   Now that we had chosen where the text is aligned, it was time to maintain this throughout the whole app.

4. **Consistency**

For example, in *Figure 6.1*, there's an equal amount of space between our three main categories (header, most played game, and last played games), and there's an equal amount of space between the displayed games under the **Last played games** header. So, we picked two different sizes, we gave them meaning, and then we used them whenever they were needed. Imagine having one game title just 2px off the bottom; maybe you wouldn't immediately notice it, but you'd feel like something was off.

The same thing can be said about colors. We chose three main colors, gave them meaning, and then maintained consistency with them. This will apply to other screens as well—the same margins, colors, and alignment. This is how we make sure our screen will never look odd to someone.

After I finish a creative process, I take a look at the before and after and try to judge for myself if one version is better than the other. Another good judge might be a relative or a friend, so don't be afraid to share your work with other people to see what they think.

My initial design will always be different than what I'll actually implement, and that's because in my opinion your first thoughts on something are always influenced by whatever is happening around you at that specific moment. It's better to take a step back and let all the information surround you. After that, you can judge your work more accurately.

Once we've settled on a design, we should be able to maintain the rules over the entire application. This is called being consistent, and it's one of the best pieces of advice I've ever received.

Maintaining consistency in your app

Consistency is something that can be extremely helpful, whether we're talking about design or our personal life. I've learned that being consistent is the key to a healthy, successful life. Being consistent is what gets me from point A to point B, and I do believe that this gets applied to every aspect of life. Being consistent is about the experience.

Slowly progressing from the main screen to the last screen of our mobile app is an experience that needs to be enjoyed by our users. This experience can only be enjoyed by avoiding confusion and reducing the amount of learning for the user.

Let's see how we can achieve consistency in our design and the appropriate way of handling issues of consistency.

Device UI guidelines and behaviors

iOS and Android have different UIs and different usability guidelines. It'd be good for you to get familiar with them. By identifying the differences between the platforms, we can ensure that our app works and functions correctly on each specific platform. Even though the designs have to be similar, there are differences in how users actually use those platforms and, because of that, you want to make sure your app will not make your users learn a different usage pattern.

Meaning

There are certain aspects of an app that we just don't want to be changed. Imagine having a blue **Submit** button for the checkout process and then a red **Submit** button for a registration form. This will create confusion.

Once we give meaning to our colors and buttons, it's important to keep the same meaning whatever screen or platform our user is using. If you're coming from the web developing industry, you'll probably know about Bootstrap. **Bootstrap** is a UI library created by Twitter that comes packed with colors, **Cascading Style Sheets** (**CSS**) classes, and design guidelines for the web. For example, they've identified a shade of blue as the color for information. This is how they maintain consistency.

Another good example can be the fact that in the screen we developed in *Chapter 4, Your First Cross-Platform App*, I chose a margin of 32px between the edge of the screen and the main content. If we were to develop another screen, we'd have to maintain the same constraints.

Language

I'm pretty sure we all know the meaning of the words inbox, submit, spam, and delete. These words are universally accepted and known by all app users. Changing words just for the sake of it requires users to develop another layer of understanding and learn these new words. For the sake of being consistent, we'll make sure all these words will have the same meaning in our apps.

Of course, the list of words goes much further than this, but a good rule of thumb is to ask yourself the following question: "Did I ever see this word or icon used in a different context in another app?" If the answer is "yes," you might want to rethink the way you're designing your application, or at least the language aspect of it.

Having discussed all these guidelines for consistency and clean design, I think we should explore different software products that might help you in designing your perfect mobile app. As we all know, there's Adobe Photoshop, which is a heavily used product in almost every aspect of design, whether it's web, mobile, or pixel art for your indie game. But we won't go into why Photoshop is as big as it is because there are other products we can use that are a lot simpler to learn and less expensive.

Figma

Figma is a tool that is compatible with almost every browser. This makes it a unique design tool as it's browser-based. You don't need to worry about installing its latest version or having to deal with compatibility issues or version issues. It is also a collaborative tool, so you can join the team designing your project.

The price is free but for better features, there's a monthly subscription payment. This is a pretty great tool and a lot of people enjoy using it.

Adobe XD

Adobe Experience Design (**XD**) is a direct competitor of Sketch. Because Sketch only works on macOS, XD is an alternative for Windows users. Of course, it also works on macOS just as well as it does on Windows. It's really fast and easy to use for a beginner. It has all the features Sketch has, such as wireframing, prototyping, and much more.

It's a free tool, but it also works on a subscription-based model for company use.

Sketch

Sketch is a really lightweight UI/UX design tool for designers. It is viewed as an industry-standard tool for prototyping, and once you're looking for more in-depth tutorials regarding design, you'll see Sketch popping out more than you'd want. It is really similar to Photoshop, but its focus is on graphic design.

Sketch has a price of **US Dollars** (**USD**) 99 right now and a free trial of 30 days. I highly recommend trying this tool as it's the standard tool used throughout the industry.

I'd have to say that my personal favorite was Sketch, but after having to design on Windows as well, I started playing around with Adobe XD. Now, I'm using XD for everything. I've even used it for the screens and example images found in this book. I love the way you can prototype on it, and I'd totally recommend trying all of them before picking your favorite tool.

Summary

This chapter was full of information regarding how to maintain a good clean design for our mobile apps. I hope that by the end of it, you've understood at least some of it because after all, we're not designers—we're programmers. I do think, though, that having at least a little bit of understanding regarding the tools other people use and some basic rules and guidelines is going to go a long way toward you becoming a better, more prepared programmer.

We've understood how to minimize user input and create great forms with a bigger completion rate for a great UX. We've also learned some rules on how to create those forms in a more logical way so that we never confuse our users.

After that, we're learned about how to declutter our designs so that they look like they have more breathing room and we've seen my creative process from the first draft for a screen design until the final result.

After learning about consistency and what exactly that is, we've explored the main ideas of each design tool so that you are able to choose the one that suits you the best.

I hope you're thrilled about the next chapter because we're getting closer to understanding and creating actual cool little applications. We'll start learning about the state of our application and how we can use it to dynamically change information throughout our app.

7
Exploring the State of Our App

After going through so many different ideas about how to build an app, it's time to get to set one of the final stones for our house's foundation. In this chapter, we'll understand what state is, and most importantly, how state works inside a React application.

We'll go through a basic definition of what exactly state is and how it was traditionally used in React applications. We'll also learn about some new modern ways of using state and how exactly they work. We'll have to decide on our own which one is the best fit to be used in our specific case, but of course, I'm going to give you my recommendations.

We'll then apply all our new information to a practical exercise, which is going to help us reinforce these new concepts in our brain so we can properly understand everything we cover.

After the practical exercise, it'll be time to look at some different hooks and what exactly they are. We'll learn about the differences about using state in a class component and how hooks can help us write less code. We'll also learn about another hook, which deals with lifecycle functions. All this will help us move forward with our studying and is imperative for us to know before creating more complex applications with React Native and Galio.

The following topics will be covered in this chapter:

- What is state?
- Leveling up our screens
- Other hooks and why they're relevant

Technical requirements

You can check out this chapter's code by going to GitHub at `https://github.com/PacktPublishing/Lightning-Fast-Mobile-App-Development-with-Galio`. You'll find a folder called `Chapter 07`, which contains all the code we've written for this chapter. In order to use that project, please follow the instructions found in the `README.md` file.

What is state?

Now that we've got to this point, it's imperative for us moving forward to understand what state is and how it works inside a React component. Once we learn this, we'll be fully capable of using React to the best of our abilities. This unlocks the missing link we've had until now, more exactly, it will unlock the key to making our mobile apps more dynamic.

We learned about *props* in *Chapter 3, The Correct Mindset*. It's the technique we use to pass data from one component to another. Think of props as the first level of a component. We need to level up our component creation skills so the most logical step now before going into any practical challenges is to learn about state.

Traditionally, in order to be able to use state inside our components, we had to use *class components*. Later versions of React, we got introduced to being able to use state even in functional components with something called *hooks*. We'll discuss hooks more right after we learn the basics of state, and for that, we have to start with class components.

What is a class, though? A class is a template for creating an object. Objects are usually used in **OOP – Object-Oriented Programming**. Although JavaScript is not a class-based object-oriented language, it still has ways of using OOP.

Let's look at how a class is created in JavaScript and what exactly it requires to function properly inside a React/React Native project:

```
class App extends React.Component {
    render() {
        return (
            <div>
                <h1>Hi everybody!</h1>
            </div>
        );
    }
}
```

Figure 7.1 – Code for a class component in React

This is really similar to a function but it does not have any *parameters* and also, we can see that extends word there. The keyword extends is basically used to let the class know it should *inherit* properties from another class; in this case, the other class is React. Component. All classes need to inherit from React.Component so the class can get used as a React component.

We also see the render() function. This function is required by a React component. It's the place where we write all our JSX. Now, there's another function we're supposed to use. It's the function that gets called when a new object is created using the class.

Now that we've been through how to create a class, it's time to finally go into state. Let's see how we can add *state* to our App class component. For that, we need to create another function inside our App class called constructor():

```
class App extends React.Component {
    constructor(props) {
        super(props);

        this.state = {
            age: 24
        }
    }

    render() {
        return (
            <div>
                <h1>Hi everybody!</h1>
            </div>
        );
    }
}
```

Figure 7.2 – The constructor function added to our class

> **Important note**
>
> In class-based OOP, a constructor is a special type of function that is called whenever we're creating an object. It often accepts arguments to custom-initialize a new object in any way we want to.

As you can see, this function accepts one argument, `props`, which enables us to use the props this component might receive. The `super()` function inside the constructor is a keyword used to access and call functions on an object's parent. This function must be used before the `this` keyword is used.

As we can see, our state variable has a `this` keyword in front of it. This keyword refers to the object it belongs to. It basically refers to the fact that the `state` variable is linked to *this* object only so you can't really access it directly from another object.

Now let's see how we can use it inside our `render` function. This is the exact same way we'd use `props`:

```jsx
class App extends React.Component {
  constructor(props) {
    super(props);

    this.state = {
      age: 24,
    };
  }

  render() {
    return (
      <div>
        <h1>Hi everybody! I'm {this.state.age} years old!</h1>
      </div>
    );
  }
}
```

Figure 7.3 – State used in the render function

As we can see, we still use the `this` keyword in order to make sure that the `state` variable refers to this specific object when it is rendered. Now the message that should appear onscreen is **Hi everybody! I'm 24 years old!**.

This is really similar to the `props` we've been using, but what exactly is different?

The actual difference is that `state` is local while `props` is something we transfer from one component to another. Another difference is that because `state` is a local variable, we can change it inside our component and the only thing that has to render is that specific component. The thing with `props` is that once we update a prop, all the children that are using that prop need to be re-rendered and that puts some **stress** on our app.

What is state?

In computer science, a system is called stateful as long as it is designed to remember previous information. The remembered information is called the state of the system.

This is **not** to say that `state` is better than `props`. They all have their purpose and you'll use all of these concepts when building an app. Sometimes, you need state and we'll look at some examples using both so we can better understand how exactly they work.

So how do we change this variable? Some of you might naturally think "Hey, this is easy – just change the variable as you'd normally do" and you're going to try something like this:

```
state.age = 54;
```

But this won't really work. You could try doing it inside your component but you won't see any differences. The state will remain 24 on your screen and the component will not re-render. React state should be treated as immutable. In the programming world, an immutable object is an object that cannot be modified after it has been created.

We actually have a React function implemented for us called `setState()`. This helps us replace the state with another state so we don't really modify the main variable; we actually replace the variable with another variable.

So, if we wanted to change the age, we would need to write something like this:

```
this.setState({ age: 54 });
```

Now, this seems fairly easy, but where exactly do we change the state? Well, there are lots of places where we can change the state but that depends on your app and how exactly you want it to work. Let's say we want to change the age right when the component is getting rendered on the screen. React gives us certain functions for our class component called *lifecycle functions*.

These functions are called at specific moments in a component's life. We'll discuss two of them: `componentDidMount()` and `componentWillUnmount()`.

These represent exactly what the names suggest. The first one gets called once our component has already mounted (rendered) to the screen. The second one is called once the component has to be removed from the screen. So, we have these moments in a component's life where we can insert code to make sure the components behave as we expect them to behave.

Obviously, if we wanted to change the age once the component gets rendered, we'd have to use the `componentDidMount()` function:

```
class App extends React.Component {
  constructor(props) {
    super(props);

    this.state = {
      age: 24,
    };
  }

  componentDidMount() {
    this.setState({ age: 54 });
  }

  render() {
    return (
      <div>
        <h1>Hi everybody! I'm {this.state.age} years old!</h1>
      </div>
    );
  }
}
```

Figure 7.4 – Using componentDidMount inside our class component

Now when we open up our app, we'll see **Hi everybody! I'm 54 years old!**. But the state actually was 24 at the start of rendering, and once it rendered, the state changed to 54. So, this is really cool, we have so many different new functions and properties. I'd totally recommend you read more about how a class works inside JavaScript if there's anything you feel you don't really understand. You can do that by visiting Mozilla's website, which is full of interesting information about JavaScript: `https://developer.mozilla.org/en-US/docs/Web/JavaScript/Reference/Classes`. Just so you know, a lot of people have problems or feel confused about how the `this` keyword works and how exactly state works. I feel like this confusion always clears once you learn a lot more about how JavaScript actually works.

Now let's use what we've learned so far and apply it to a cool little practical exercise. We'll now begin using state in order to make our screens look more dynamic instead of just our usual static screens.

Leveling up our screens

Let's see what kind of app we're going to create. I was thinking we could have a screen where it shows our current age down to months, days, hours, and minutes. I mean, that's pretty cool, right? Whenever someone asks you your age, you'll be able to just take your phone out of your pocket and show them the screen you've created. Let's get started:

1. Let's open up our terminal and create a new `expo` managed project just like we always do, using the following command:

    ```
    expo init RealAge
    ```

    Now let's open up our project and start writing some code!

2. Let's go straight to the `App.js` file and delete everything in there besides the imports and StyleSheet. I always leave the StyleSheet because I'm a fan of centered text.

3. Now let's rewrite the `App` component as a `class` component.

```
import { StatusBar } from "expo-status-bar";
import React from "react";
import { StyleSheet, Text, View } from "react-native";

class App extends React.Component {
  render() {
    return (
      <View style={styles.container}>
        <StatusBar hidden />
        <Text>My real age is: </Text>
      </View>
    );
  }
}

const styles = StyleSheet.create({
  container: {
    flex: 1,
    backgroundColor: "#fff",
    alignItems: "center",
    justifyContent: "center",
  },
});

export default App;
```

Figure 7.5 – App.js rewritten as a class component

4. Now let's open up our Expo app by using the following command inside
    our terminal:

```
expo r -c
```

I always use this command as it clears out the cache. So, this way, I make sure the
cache will never interfere with my changes.

5. Now that the Expo server is open, just as we learned, open up the simulator of your
    choice. You should be able to see the text `My real age is:` on the screen once
    your app opens up.

6. Now let's integrate our age as a state inside the `App` class component.

Just as we've seen before, we need to write our `constructor()` function above
everything else inside our class component. *Don't forget* about the `super(props)`
line – that one's important! We'll then create our state inside our `constructor`
function:

```
class App extends React.Component {
  constructor(props) {
    super(props);

    this.state = {
      age: {
        years: 0,
        month: 0,
        days: 0,
        hours: 0,
        minutes: 0,
        seconds: 0,
      },
    };
  }
```

Figure 7.6 – constructor function with our newly created state

I've already mentioned we're going to display our age in terms of years, months, and days,
all the way to seconds, so I've put an object filled with zeros in there just as a placeholder.
It could've really been anything in there as we're going to change it after some quick math.

Now let's dive right into how we're going to calculate the age. For this little trick,
we're going to use the `Date()` object inside JavaScript. Even though this object might
seem a little bit confusing for some people, after you learn more about time zones, it really
becomes just another object to play with. Don't sweat it, we're not going to go that deep
into dates with JavaScript as we have better stuff to learn about.

So, we're going to create a new function called `getAge()`, which is going to receive your birthday date. This function will take the current time and will subtract from it the date of your birth. All of this will be done in milliseconds. After that, we're going to take the result and create a new `Date` object with it. From that new object, we're going to extract all the information about how old we are.

At the end of it, we're going to use `setState` to create a new state with all the information we've calculated from our `Date` objects:

```
getAge(dob) {
  var diff_ms = Date.now() - dob.getTime();
  var age_df = new Date(diff_ms);

  this.setState({
      age: {
          years: age_df.getUTCFullYear() - 1970,
          month: age_df.getUTCMonth(),
          days: age_df.getUTCDate() - 1,
          hours: age_df.getUTCHours(),
          minutes: age_df.getUTCMinutes(),
          seconds: age_df.getUTCSeconds()
      }
  });
}
```

Figure 7.7 – Our function to calculate our current age

Now, you're probably wondering why we subtracted `1970` for years and `1` for days. Oh well, as I was saying, the `Date` object is a little bit weird. We had to subtract `1970` because UTC time starts at 1970, so in order to be sure we're getting our correct year value, that had to disappear from our equation. As for the value for days, this might have something to do with the fact that I really wanted to make sure that time zones would be taken into consideration and my time zone needed that `-1`. The thing is, even if we get 1 day off, the important thing is to see this thing really work.

Now that we have the function, and we're using the `setState` function to correctly change the state, it's time to call this function from somewhere. As you know, a normal function won't just call itself (even though there are functions out there that can do that).

So, let's do the same thing we did before – let's call our function in `componentDidMount()` like this:

```
this.getAge(new Date("June 29, 1996 03:32 UTC+2"));
```

As you can see, I've used the keyword `this` to make sure our object knows we're referring to its function `getAge`. I've also used my own birthday inside the function but you may use your own birthday to make this even more personal.

Our job is not done! Let's get to our `render` function and make some modifications so we can display everything properly:

```
render() {
  const { years, month, days, hours, minutes, seconds } = this.state.age;
  return (
    <View style={styles.container}>
      <StatusBar hidden />
      <Text>My real age is: </Text>
      <Text>{years} years</Text>
      <Text>{month} months</Text>
      <Text>{days} days</Text>
      <Text>{hours} hours</Text>
      <Text>{minutes} minutes</Text>
      <Text>{seconds} seconds</Text>
    </View>
  );
}
```

Figure 7.8 – Our render function after we've implemented our state

The first line inside our `render` function might seem a little bit weird to some of you. It's called **object destructuring**. This is what we've already been doing with our imports. This is a really useful JavaScript feature used to extract properties from objects and even bind them to variables.

For example, we're now able to just say `years` whenever we're referring to `this.state.age.years`. It saves us writing time and it also looks a lot clearer. You'll see people destructuring variables like this all the time – it's a really cool feature!

Now that we've made sure that we're going to use all the variables inside our `state`, our `componentDidMount` is calling our `getAge` function and the `state` is set inside that function, everything is ready. Run your app and check out the result. You should be able to look at the screen and see how old you really are, down to the smallest detail.

But there's something wrong – the seconds don't refresh, so everything stays the same. You're probably thinking that I could've lied to you, but trust me I didn't. Right now, your real age is not updating because our `getAge` function is only getting called once. As we said, `componentDidMount` calls the function when the component first renders on the screen. Our component rendered, the function got called, and that's the end of the story.

We somehow have got to make that function call multiple times; I'm thinking at least once a second so we make sure our seconds are in sync with the real time. Let's do that now!

Inside our `componentDidMount` function, we'll call a cool little function called `setInterval()`. The first parameter it accepts is a function. This will be called at an interval of time. The second parameter it accepts is actually the time in milliseconds for how often to execute the function:

```
componentDidMount() {
  this.tick = setInterval(() => {
    this.getAge(new Date("June 29, 1996 03:32 UTC+2"));
  }, 1000);
}
```

Figure 7.9 – componentDidMount with our setInterval function

Now we've created this interval at which our `getAge()` function is called. It's a good practice to stop the interval when we don't really need it to work anymore. The question popping right now in your mind is probably "When don't we need it run?". Well… That's usually subjective but in our specific case, the answer is at the end of our component's life.

Remember we said there's another lifecycle function called `componentWillUnmount()`? Well, that's exactly where we're going to stop this function:

```
componentWillUnmount() {
  clearInterval(this.tick);
}
```

Figure 7.10 – The componentWillUnmount function used in our class component

Now that we've done this, our app should be ready to display our current age correctly. Save everything, refresh the simulator, and check it out! Your real age is now displaying right on the screen. Don't let those numbers ruin the day for you though – we're all as young as we feel!

Now that we've seen how state behaves in a `class` component, which is a bit more of a traditional use of state, it's time to see other ways of using state. In the recent past, React blessed us with some cool little things called **hooks**. Let's learn more about them, how exactly they differ from our traditional state, and what new features they bring to the table.

Other hooks and why they're relevant

The main problem with state is the fact that the only way we can use it is in a `class` component. Class components are generally seen as a bit ugly and hard to learn for some beginners, so the React team tried creating something new that promised to solve the problems beginners and advanced users could have gotten into while using a class component with the traditional use of state. This is how **hooks** were born.

Hooks were introduced in React v16.8 and React Native v0.59. They basically let you use state and other React features without writing a class.

So, what exactly does that mean for us? Let's look at an example of how state is written with our new hooks feature:

```
import React, { useState } from 'react';
import { Text, View, Button } from "react-native";

export default function Example() {
  const [count, setCount] = useState(0);

  return (
    <View style={styles.container}>
      <Text>You've pressed the button ${count} times.</Text>
      <Button title="Click me" onPress={() => setCount(count + 1)}/>
    </View>
  );
}
```

Figure 7.11 – Example of using hooks

Woah! So, what do we have here? Is this really the same state feature we've been using so far? Yes, it is. If you were to copy this code into a fresh new project, you'd see that once you started up your app, every time you pressed that button, the number would keep updating from 0 to however many times you pressed it.

Let's see what exactly we wrote here.

As you can see, we've created a function called `Example`. The name doesn't really matter as long as it's not your main function, which should always be called `App`. A function looks much cleaner than a class and it's obviously a lot easier to write.

Then we've defined two variables inside our function using the hook `useState()`. How exactly does that work?

```
const [count, setCount] = useState(0);
```

In this example, useState is a hook. We call this method inside a function component in order to add local state to our component. This function returns a pair: the *current* state value – count, and a function that lets you update that value - setCount. The setCount function is pretty similar to the this.setState function in a class, except it doesn't merge the old and new state together.

The only argument useState accepts is the initial state given to our count variable. Remember that our this.state variable had to be an object and everything was inside that object. count doesn't have to be an object even though it could be if you want it to.

Now let's see a straight comparison between using this.state and the useState hook. We'll see the same state written with both of these features so we can have a clear way of comparing the two.

First, we'll take a look at this.state. We'll imagine having an app that needs to have some information regarding a user, some comments that their friends have left on the user's profile, but also the number of likes this profile has:

```
this.state = {
  userInfo: {
    age: 24,
    name: 'Alin',
  },
  comments: ['OMG', 'cute', 'old picture'],
  likes: 34
}
```

Figure 7.12 – State object as written in a class component

This is pretty easy to understand, right? Our state has the following values: userInfo – an object, comments – an array of strings, and likes – a number. Let's see how the same thing would look using *hooks*:

```
const [userInfo, setUserInfo] = useState({ age: 24, name: 'Alin' });
const [comments, setComments] = useState(['OMG', 'cute', 'old picture']);
const [likes, setLikes] = useState(34);
```

Figure 7.13 – Our state written in a functional component

This is the exact same thing but we've been using the useState hook. Everything has the exact same values as the previous example but the difference is in the fact that our state is not living in a single object.

Now, let's say, for example, we want to change the number of likes. Maybe someone clicked the like button and we want to update the number displayed on the screen. Let's see how we would change it in a class component:

```
this.setState(prevState => {
  return {
    ...prevState,
    likes: prevState.likes + 1
  };
});
```

Figure 7.14 – Changing the state in a class component

This looks complicated, right? On top of that, there's a bunch of new things compared to the usual `setState()` function we've been using until now. The thing is, because we need to update the state just for the likes value, we used something called *previous state*. That's where `prevState` comes from. Once you need to change the state based on the previous state, as we need to do here because we need to increment the number of likes, it's imperative to pass to `this.setState` a function as an argument. This provides us with a snapshot (`prevState`) of the previous state. We've been using the short version until now because we didn't need to update it based on the previous state.

Now let's see how the same thing would look if we'd been using hooks:

```
setLikes(likes + 1);
```

Figure 7.15 – Changing the state in a functional component

This is obviously a lot cleaner and easier. We know we want to change only the likes, so we're using `setLikes`. Here, we can take the `likes` state and just increment it by `1`.

As you can see, hooks make our life a lot easier. They're really simple to use and require a lot less writing.

Now, the thing is, if we were to take the app we created before going into **hooks**, the one that displays our real age, how exactly would we be able to call the `setInterval` function because the lifecycle functions – `componentDidMount` or `componentWillUnmount` – are only available in a class component.

We're in luck because the React team provides us with a lot more hooks for us to use besides `setState`. First, let's see what exactly a hook is.

As we know, React is all about code reusability. Right now, we can write simple functions and call them whenever we need to calculate something or even write components in order to reuse them in any part of our application, but the problem with components is the fact that they have to render some UI. This makes components kind of inconvenient. The React team got the hooks idea because they wanted to be able to share complex logic without having to render some sort of UI. Hooks let you use React features from a function with just a simple function call. The hooks we've been provided with cover the most important parts of React: state, lifecycle, and context.

So, let's see what type of hook we could use instead of the `componentDidMount` function.

useEffect

The `useEffect` hook enables us to use side effects from a function component. What are *side effects*? For example, *data fetching* or *subscriptions* are side effects. They're called that because they can affect other components and can't be done during rendering.

Usually, those operations are performed with lifecycle functions in a class component. You can think of `useEffect` like all those lifecycle functions all combined in a single function. Just like `useState`, `useEffect` may be used multiple times inside the same functional component.

By using this hook, you'll basically tell React that your component needs to do something after rendering. React will remember the function you passed and call it later after performing all the updates. `useEffect` runs after every render. So basically, it runs after the first render and after every update your component does.

Okay, so what about `componentWillUnmount`? How can we make sure that our function will only work when it's time to remove the component? `useEffect` is enough for this and we don't need another hook. If we do return a function from our effect, React will make sure to call that function once our component is unmounted.

Hooks are a really big part of React and they require lots of explaining, and I feel you'd get the most out of just reading the documentation. There are other hooks out there, for example, `useMemo`, `useRef`, and `useReducer`. So, reading the documentation is a lifesaver for all programmers, especially because you'll find in there lots of really cool information that I can guarantee you won't find in any book. When learning a new technology, your first step should be the documentation and then researching other ways that are more specific and more to the point about what you're really trying to study. Just like this book, we're here to learn how to build some React Native cross-platform applications, so let's move on and we'll explain more about hooks when we get to the point of using them in the next chapter.

Summary

This chapter has gone over most of the information about state required for us to move forward. By now, we should be able to understand how state works both in a class component and a functional component.

After learning about state and what exactly state is, we learned about some lifecycle functions and how exactly they work. Learning about this is really important because we've now understood that a component lives through different stages and at different points, we're able to interfere with some JavaScript code.

This whole adventure gave us an idea, the real age app. We're now able to create an app with dynamic numbers that change over time. We've learned how to implement everything we've learned so far about state and create an awesome idea displaying our age.

Because class components look a bit like there's too much code to write, we started learning about hooks. After a careful analysis of how exactly they're different, we learned about a hook called `useEffect`.

Learning all this will be really beneficial in the long run, especially in the following chapters when it's all about practical challenges, where we'll learn lots of tricks and create many different types of React Native applications.

8
Creating Your Own Custom Components

After getting through all these lessons, we're now ready for more practical challenges, which are going to get us ready for creating fully-fledged React Native applications. We've touched upon all the basic and a bit more advanced information, so we're prepared to take on more difficult challenges.

In this chapter, we're going through four different exercises. The first one is going to be a simple exercise where we're going to use Galio's components to create new ones that'll fit our imaginary app. Doing this will once again prove to us how helpful Galio can be for almost all of our programming needs.

After that, we're going to create our own profile card. This exercise will mostly focus on layout and styling, as I feel this is a really important part of any app creation. Learning this will get us one step closer to creating the app of our dreams because nowadays, almost every app has a profile screen or card included somewhere.

The next exercise will deal with controlled inputs. On top of creating a simple registration form and styling it to the best of our abilities, we'll also understand how state is necessary when working with inputs or forms in general.

The final challenge for us will be creating an e-commerce card. This will serve as proof of the fact that almost anything can be created by figuring out how it's similar to something you've already created. This is the moment where we can understand that having experience in a field will definitely help in another field. No experience is useless; everything helps us grow as a better human being overall.

The following topics will be covered in this chapter:

- Creating your own component!
- Creating your own profile card
- Creating your own register form
- Building your e-commerce cards

Technical requirements

You can check out this chapter's code by going to GitHub at `https://github.com/PacktPublishing/Lightning-Fast-Mobile-App-Development-with-Galio`. You'll find a folder called `Chapter 08` that contains all the code we've written inside this chapter. In order to use that project, please follow the instructions found in the `README.md` file.

Creating your own component!

Now that we've gotten through all the basic knowledge about how React and React Native work, it's time to put our skills to the test by creating a lot of different components. Worry not—we're also going to create a bigger and more complex app. But as you know, a React application is formed from lots of different components, so by creating components, we're actually getting ready to create apps.

I was thinking that for our first component, we should start with a news card. This would usually go straight to the news feed—we'd use multiple components like this with different text if we were to create a news app. So, how do we start?

Just like we usually do, create an app with the following command:

```
expo init chapter08
```

We're going to use the same app for all our exercises in this chapter because it's a lot easier than creating a project for each of them. So, right after the project has been created, let's open it up and then open our App.js file.

Now, we're going to create a new components folder inside our root folder. Here, we're going to start developing our own components. At the end of this chapter, you should have four files inside this folder.

Because we're going to use Galio for creating our component's layout, we should install it now via the terminal. Remember the command we're using for installing external packages? We're going to use the following command:

```
npm i galio-framework
```

Now, let's create a new file called NewsCard.js inside our components folder. Because we're creating a news feed type of component, we need to think of what exactly we would need to use inside this component.

We know for sure that we need StyleSheet for the styles and the Block component from **Galio**. But we also need a Text component for rendering the text and an Icon component, so that we'll be able to have some sort of icon. I feel that every post should also have an avatar, so an Image component is also required.

So, our imports should now look like this:

```
import React from 'react';
import { StyleSheet, Image } from 'react-native';
import { Block, Text, Icon } from 'galio-framework';
```

Figure 8.1 – Imports used for our NewsCard component

Now that we know what we're going to use inside our component, let's start building it piece by piece. We'll start by creating a functional component named NewsCard. This function is going to return, for now, just a Block element and a Text element in order to have something to be rendered.

We'll also create a styles object at the end of the file. Remember how we were supposed to do that? Nice! Let's create a style called card for our main Block component. For the styling, I was thinking of adding something new that we haven't discussed until now: shadows.

Shadows are really not that hard to use but I feel that some people might not really understand how those work. After adding the style, let's take a look at what our component looks like up to now:

```
export default function NewsCard(props) {

    return (
        <Block style={styles.card}>
            <Text>News Card</Text>
        </Block>
    );
}

const styles = StyleSheet.create({
    card: {
        backgroundColor: '#fff',
        padding: 8,
        width: '80%',
        borderRadius: 6,
        shadowColor: '#000',
        shadowOffset: {
            width: 0,
            height: 2
        },
        shadowOpacity: 0.23,
        shadowRadius: 4,
        elevation: 7
    },
});
```

Figure 8.2 – The beginning of our first component

So, everything should look fairly simple to understand at this point. The shadows here are the only topic we haven't really explored, but the styling should be self-explanatory. First, we have shadowColor, to which we've assigned #000, which is black. Then we have shadowOffset, which is telling our shadow how to fall down from the object we've been assigning it to. If things still seem a bit confusing, we should think about width and height values like this: width is the *x* axis and height is the *y* axis. Saying width: 0 means that our shadow is expected to fall down straight to the ground underneath the object, but combining it with height: 2 will tell our shadow to drop down 2 **pixels (px)** from the center. Then, we have shadowOpacity, which does what you'd expect it to do: it calculates the opacity of our shadow. You've probably noticed elevation; this is what you're using to set the shadow for Android devices, and it only is supported on Android 5.0+.

Now that we've set the basis of our new component, let's import it to App.js so that we can see our changes in real time. So, let's open up the file and delete everything inside the main function besides the main View component. Keep the styles—I love having everything centered.

Now, let's import our newly created component and render it on the screen. We'll go right below our main imports and write the following code:

```
import NewsCard from './components/NewsCard';
```

Now that we've imported the component, place it inside the View component like this: <NewsCard />. Start up the Expo server, open the simulator, and you should be able to see the card with the text **News Card** on it. Great! Now, we can work on it—save the file and see the changes in real time.

Eventually, we'll add each component we're creating in our App.js file. This should be a pretty easy workflow to test our components visually.

Now, let's go back to our NewsCard.js file and start creating the basic layout.

We'll start by arranging the layout with Block components, so we'll use two of these. The first one is for the header of our card, which will contain the bookmark icon to the far right of the card, and to the left side, we'll have the avatar and information about the author. The second one is for the title of the news article and a summary of the text. Let's see how that looks right now by putting it into practice, as follows:

```
return (
    <Block style={styles.card}>
        <Block row space="between">
            <Block row>
                <Text>Avatar</Text>
                <Block style={{ paddingLeft: 8 }}>
                    <Text size={14}>{props.author}</Text>
                    <Text size={10}>{props.date}</Text>
                </Block>
            </Block>
            <Icon name="bookmark" family="feather" color={'rgb(42,92,250)'} size={20} />
        </Block>
        <Block style={{ paddingTop: 8 }}>
            <Text size={16}>{props.title}</Text>
            <Text size={14}>{props.summary}</Text>
        </Block>
    </Block>
);
```

Figure 8.3 – Coding the basic layout

So, as far as you can see, for `title`, `summary`, `author`, and `date`, we'll be using `props`. As for `Avatar`, right now we'll be using a `Text` component as a placeholder. So, let's save and move to our `App.js` file to finish sending all the props back to our `NewsCard` component, as follows:

```
export default function App() {
  return (
    <View style={styles.container}>
      <NewsCard
        author="Richard"
        date="10 minutes ago"
        title="Cleaning your computer"
        summary="Lorem ipsum dolor sit amet, consectetur adipiscing elit."
      />
    </View>
  );
}
```

Figure 8.4 – App.js file with props completed for our NewsCard component

Right—now, we're going to save the `App.js` file and switch to our simulator. We should be able to see our `NewsCard` component taking form. There's a title, a summary, a date, and even an author. Yeah—I've used `lorem ipsum` for the summary because it's easier and quicker than actually creating a summary text for our dummy component. We could even start a news feed with our components. But right now, let's go back to our `NewsCard` component and add the stuff that we're still missing.

For sure, there's a need for us to replace the placeholder we've been using with an actual `Image` component. So, let's replace that text with the following line:

```
<Image style={styles.avatar} source={{uri: props.avatar}}/>
```

You might remember that an image needs to have some styling attributed to it in order to render. Let's go to the `styles` object and do all the styling we need for our image. I was thinking of having a `width` and `height` value of `30px` and a `borderRadius` value of `15px`.

Now, the only thing we're missing is going back to our `App.js` file and adding the `avatar` prop to our component. Search an image online and paste the link in there. Now, refresh everything, and congratulations—we have an image rendered!

I'd say that right now, the only thing we're missing is to add some colors to the text, but I'll let you do that on your own. If you haven't coded this at the same time as me, worry not—just go to GitHub and search for the `Chapter 08` folder. This is going to have all the code we've done until now, and you'll also see how I've colored the text. I've also destructured the `props` object.

Now, let's see how this looks on my simulator so that you can make sure that once you clone the repository on GitHub, things look the same way we've been describing them. You can see the result here:

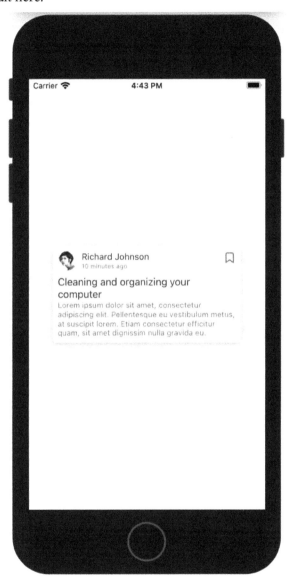

Figure 8.5 – Simulator displaying our finished component

This looks pretty good, right? I'll let you use this component as much as you want in your future apps, so don't shy away from reusing your components. The next one should be cooler, so let's move on and start building our first profile card.

Creating your own profile card

A profile card is something any user needs to see inside an app with a user system. So, I was thinking of creating a simple profile card that is going to display some basic information for our users. The main elements that I feel should be displayed are a profile picture and the user's name, email, and phone number.

This will serve us great purpose in an app where maybe we have a list of phone contacts and we want to see each contact separately. Now, let's start creating our profile card component.

Go ahead and create a new file in our `components` folder called `ProfileCard.js`. Now, as you read earlier, I've stated which elements this component will be composed of. Based on that, let's think of what type of imports we need.

You guessed it! The same imports we've been using in our last component. Now that we're sure of what type of imports we need to have, let's write a basic function so that we can get something on the screen to see while we start working on the component.

As you can see in the component's filename, our main `Block` component should be a card, so let's apply the same styling that we applied to our last component. We'll change the background color and some values, but this `style` object should be mostly the same as the last one.

Let's take a look at what we've been writing until now:

```
import React from 'react';
import { StyleSheet, Image } from 'react-native';
import { Block, Text, Icon } from 'galio-framework';

export default function ProfileCard(props) {

    return (
        <Block style={styles.card}>

        </Block>
    );
}

const styles = StyleSheet.create({
    card: {
        backgroundColor: "#006d77",
        width: "80%",
        borderRadius: 14,
        paddingHorizontal: 16,
        paddingVertical: 32,
        shadowColor: '#83c5be',
        shadowOffset: {
          width: 0,
          height: 2
        },
        shadowOpacity: 1,
        shadowRadius: 8,
        elevation: 7
    },
});
```

Figure 8.6 – The start of our ProfileCard component

Things look really similar, right? There are some values changed, but this is because I feel different colors might suit this card better. It should be of the same width as our previous component based on the "80%" value we've assigned to the width property.

Now, let's go to our App.js file and comment out our <NewsCard /> component and import our new component, just like we've done before.

Now, we should be able to see this small card with no content on our simulator's screen. Let's go back to our card and continue adding the rest of the layout.

We should have an icon on the left side of our component, something that the user might want to press in order to modify the contents of our component. We'll not create the functionality yet, but having an icon there pointing to this functionality should be good enough for us.

Below this icon, I feel we should have an avatar and the contact's name centered on the card.

Right below these, the phone number and email should be available for us to see. Between those two, I was thinking of having a line dividing the information. Why? It just looks better, in my opinion. So, let's move on to the next step and add all the basic components we need for this type of layout, as follows:

```
export default function ProfileCard(props) {

    return (
        <Block style={styles.card}>
            <Icon name="gear" family="font-awesome" color="#edf6f9" size={21}/>
            <Block center>
                <Image style={styles.avatar} source={{ uri: props.avatar}} />
                <Text size={18} bold>{props.name}</Text>
            </Block>
            <Block row space="between">
                <Text bold>Phone</Text>
                <Text>{props.phone}</Text>
            </Block>
            <Block />
            <Block row space="between">
                <Text bold>Email</Text>
                <Text>{props.email}</Text>
            </Block>
        </Block>
    );
}
```

Figure 8.7 – Basic layout for our ProfileCard component

This sort of layout is really easy to prototype with Galio. As you can see, all we're using are `Block` components and we can already center, create rows, and define the spaces that each component needs. Again, we're using `props` because your job right now is to go back to `App.js` and pass down the `props` to our component so that it can render with more information.

Done? Great! You might be wondering now what's up with that `<Block />` component between the two rows we've created. Well, that'll act as a divider. So, let's write the styling for it and for our `avatar` image. At this point, you can even go ahead and add colors for each `Text` component so that you can make this look a lot more interesting. I'd probably use white for the text, but any color would work as long as you're happy with it. Let's check out how our styling looks, as follows:

```
avatar: {
    width: 100,
    height: 100,
    borderRadius: 50,
    marginBottom: 16
},
divider: {
    height: StyleSheet.hairlineWidth,
    width: '100%',
    backgroundColor: '#edf6f9',
    marginTop: 8,
    marginBottom: 8
}
```

Figure 8.8 – Styling for our divider and avatar

Now that we've created the styling, let's dive in for a second. The divider should be a sort of white line between our `Email` and `Phone` number. So, we've used a `Block` component to create a straight line. This should make you realize in how many ways you can use a `Block` component. What's up with `hairlineWidth`, though? This is defined by React Native as the width of a thin line on the specific platform. It is mostly used for creating a division between two elements.

Now, let's save everything and see how it looks on the simulator. The output should be similar to what I have right here. Maybe you've changed some colors, but the layout should look identical:

Figure 8.9 – Final render of our component

This has been a real adventure! We've already created two different components and we're not stopping here. I hope you're having fun and following closely with some code in front of you. It's always a good idea to recreate everything from memory in 2-3 days. Just a cool little exercise you can do to make sure you're learning everything that you're reading. Now, let's move forward because the next one is going to be really cool.

Creating your own register form

Register forms are used almost in every app you can think of. You might need one, so let's see what's going on when creating a register form. This is pretty cool because, on top of creating a nice little register card, we're also going to learn something new about inputs.

Let's start how we always start—comment out the previous components from App.js and create a new file in our components folder called RegisterForm.js.

We've already created two components, so let's see whether you can start creating this on your own. The form in the following screenshot will be the final rendered version of our register form. I've chosen to let you look at it before you actually start creating it because I think you should be able to achieve a similar result on your own without my help. Of course, I'll still help you, but this is a good chance to take your time, close the book, and start creating this on your own. Check out the following screenshot and start creating on your own!

Figure 8.10 – Final rendered version of our register card

Looks pretty neat, right? This isn't really hard to create based on what we've done until now. So, now that you've taken a look at it, maybe you're already thinking about how to start working on this component. That's great! If you're still reading, that's also fine because we'll go right ahead and start creating this component.

Just as we've done up to now, we're going to start thinking of what types of imports we need to have. We don't need an image anymore, but we do need an `Input` and a `Button component`. Worry not about the icons placed inside the inputs—you can do that directly from the `Input` component. Galio makes it really easy to style and add icons inside your input.

I feel that our inputs should look something like this for this specific component:

```
import React from 'react';
import { StyleSheet } from 'react-native';
import { Block, Text, Input, Button } from 'galio-framework';
```

Figure 8.11 – Imports used for our register form

Can you already think of how we should be creating the layout for this one? We don't need any rows here because all the elements are coming straight down in a column. The only `Block` element we'll be using is the one used for creating the card itself.

Let's start by writing our main function, just like we did before. We need a `Block` component with the card styling applied to it, as shown here:

```
export default function RegisterForm() {

    return (
        <Block center style={styles.card}>

        </Block>
    );
}

const styles = StyleSheet.create({
    card: {
        backgroundColor: "#fff",
        width: "80%",
        borderRadius: 14,
        paddingHorizontal: 16,
        paddingVertical: 32,
        shadowColor: '#000',
        shadowOffset: {
            width: 0,
            height: 2
        },
        shadowOpacity: 0.15,
        shadowRadius: 6,
        elevation: 7
    }
});
```

Figure 8.12 – The beginning of our RegisterForm component

Now, let's go into our `App.js` file and comment out our previous components so that we can import our newly created component. We've done this multiple times by now, so this should be easy.

Now, let's continue with our component and quickly go through the layout. As we've already done this multiple times, this shouldn't be hard to understand.

We obviously start with a `Text` component, followed by three `Input` components and one `Button` component. So, let's write that down, as follows:

```
export default function RegisterForm() {

    return (
        <Block center style={styles.card}>
            <Text size={36} bold color="rgba(0,0,0,0.8)" style={{ marginBottom: 24 }}>app name</Text>
            <Input placeholder="Name" style={styles.input}
                icon="person"
                family="ionicons"
                iconSize={24}
                iconColor="#ffc8dd"
            />
            <Input placeholder="Email" style={styles.input}
                icon="mail"
                family="ionicons"
                iconSize={21}
                iconColor="#ffc8dd"
            />
            <Input placeholder="Password"
                password viewPass style={styles.input}
            />
            <Button color="#cdb4db" shadowless onPress={() => {}}>
                Register now
            </Button>
        </Block>
    );
}
```

Figure 8.13 – Our almost completed component

OK—so, everything is pretty much what we've always done up to now. Let's tackle the new things found here. So, on our third `Input` component, we can see two props: `password` and `viewPass`. The first one is for making sure you cannot see the password as you write; it transforms your writing into those dots we so often see whenever we're typing out our password somewhere. The second one is there to display that icon on the right, which the user can press in order to see whether there's something wrong with the password they just typed, basically transforming the dots into letters and vice versa.

Our `Button` component also has that `shadowless` prop, which does exactly what you'd think: it makes the button shadowless.

Now, here comes the interesting part. Of course, we'd like to know what the user is typing; how else are we going to verify that the information is correct or even typed the way we want it to be typed? Maybe you asked for the user's email, but what if the typed words are some random words just to break into the app without actually registering? So, there must be some way we can make sure that we know what the user has typed and verify that text once the user presses the **Register now** or **Submit** buttons.

This technique is called **controlled components**. A controlled component takes its current value through props and sends any changes through callbacks. A parent component "controls" it by handling the callback and managing its own state and then passing the new state values as props to the controlled component.

In most—or even all—cases, you should use controlled components when we're dealing with forms.

Because we're in a functional component, we'll be using **hooks** for our states. Don't forget to import the useState hook, as follows:

```
const [name, setName] = useState("");
const [email, setEmail] = useState("");
const [password, setPassword] = useState("");
```

Figure 8.14 – Hooks used inside our functional component

This is pretty easy as we've already learned about hooks and a component's state. Now, let's apply our states to our Input components, as follows:

```
<Input placeholder="Name" value={name} onChangeText={setName} style={styles.input}
    icon="person"
    family="ionicons"
    iconSize={24}
    iconColor="#ffc8dd"
/>
<Input placeholder="Email" value={email} onChangeText={setEmail} style={styles.input}
    icon="mail"
    family="ionicons"
    iconSize={21}
    iconColor="#ffc8dd"
/>
<Input placeholder="Password" value={password} onChangeText={setPassword}
    password viewPass style={styles.input}
/>
```

Figure 8.15 – State applied to our Input components

So, what exactly happens here? Once the user presses on an **input** and starts writing their name or email, for example, our onChangeText prop triggers our setName prop, which sets the name state variable to the current value of our input. This way, we're making sure that our RegisterForm component is *controlling* the *inputs* and is constantly updated with information about our input's state.

It might be somewhat hard for some people to grasp why we need it. The truth is this is how React is making sure that there won't be any errors related to our input's state, while also giving us full control and knowledge of our input's current state at all times.

Now, let's write a simple verification for our form. We need to at least have a message for our users popping up in case there's nothing written there and the user presses the **Register now** button.

So, we'll create a function called registerButton. You can name it however you want, but I called it this because it made sense to me. This function will verify the length of our input's value. Now, if we weren't having this controlled component, we wouldn't have been able to access those values via normal variables. We might have to use something called refs.

This is mostly beside the point as we're not going to learn about refs, but it's important to know that there is something called refs. Let's take a look at this registerButton function, as follows:

```
const registerButton = () => {
    if(name.length !== 0 && email.length !== 0 && password.length !== 0)
        alert(`Registration complete! Thank you, ${name}! We'll send you an email to ${email}`);
    else
        alert("You forgot to fill in the form :(.");
}
```

Figure 8.16 – Our form verification function

Now that we have this function, we just have to call it whenever the user presses the **Register now** button, so we're going to use the onPress prop on our Button component. Go ahead and apply this prop to our Button component, like this:

```
onPress={() => registerButton()}
```

Now, save and refresh the app and try it out! Pretty cool, right? There's a message popping up whenever you try to press the button with no text inside the inputs, while there's also another cool message where we're using the values written in our inputs.

This wasn't a hard component to be created but we've learned about something really cool, and that's *controlled components*. I hope this little exercise managed to successfully teach you something new that you'll be using quite often from now on whenever working with forms.

Now that we've finished with this component, let's move ahead and start working on a different component involved with e-commerce mobile applications.

Building your e-commerce cards

As you know, whenever you buy something online, there's always a basket full of the products you've chosen. Each item in that basket is usually a card with information about the price, the item's name, a picture, and the possibility to increase or decrease the number of items of the same type.

So, this is what we're going to create as well. Let's take a look at it here, as we've already become so advanced that we should now be able to think of ways of creating the functionality with only those things that we've learned up to now:

Figure 8.17 – Final rendered version of our e-commerce card

Looks pretty good, right? Honestly, it isn't even that hard to build, so we'll be moving quickly through the layout. Let's create a new file called `CommerceCard.js` inside the `components` folder.

Now, let's think about which types of imports we'd need for this one—obviously, a `Block` and `Text` component. We'll also need to import the `Icon` component because, as we can see in *Figure 8.17*, we have a minus button and a plus button there. To make those buttons clickable, we'll be using a `react-native` component called `TouchableOpacity`, so let's import that as well. On top of that, as we can all see, we also have an `Image` component. Let's see what all of our imports look like, as follows:

```
import React, {useState} from 'react';
import { StyleSheet, Image, TouchableOpacity } from 'react-native';
import { Block, Text, Icon } from 'galio-framework';
```

Figure 8.18 – The imports we'll be using for creating a CommerceCard component

We've also imported `useState` because the number will change based on which icon we're pressing. So, let's start creating our functional component now, as follows:

```
export default function CommerceCard(props) {

    return (
        <Block row style={styles.card}>
            <Image style={styles.item} source={{ uri: "https://picsum.photos/100" }}/>
            <Block style={{ marginLeft: 16 }} flex space="around">
                <Text>{props.itemName}</Text>
                <Block row space="between">
                    <Text>$ {price}</Text>
                    <Block row space="between">
                        <TouchableOpacity style={styles.buttons} onPress={() => {}}>
                            <Icon name="minus" family="entypo"/>
                        </TouchableOpacity>
                        <Text>{quantity}</Text>
                        <TouchableOpacity style={styles.buttons} onPress={() => {}}>
                            <Icon name="plus" family="entypo"/>
                        </TouchableOpacity>
                    </Block>
                </Block>
            </Block>
        </Block>
    );
}
```

Figure 8.19 – Our component's layout

That doesn't look that hard to read, right? Let's explain some of it because we've moved a bit faster right now. But this is because I feel like you've advanced quite a bit, so you should just try to challenge yourself and see whether what you think matches with the component that I've written and we've seen in *Figure 8.17*.

As far as we can see, we can have a big `Block` component containing everything and making the content inside in a row. The first element in the row is our image. After that, we have another `Block` component with the prop of `flex`, which is basically telling our component to take as much space as it can.

Inside that `Block` component, we have a `Text` component that receives the name of the item as a prop called `itemName`. We then have another `Block` component with the `row` prop applied, which will be used to separate the price and the quantity, both of which are going to be state variables.

Now, let's see what the styling looks like—trust me, the styling is a piece of cake. Here it is:

```
const styles = StyleSheet.create({
    card: {
        backgroundColor: '#bde0fe',
        padding: 8,
        width: '80%',
    },
    item: {
        width: 50,
        height: 50
    },
    buttons: {
        marginHorizontal: 4
    }
});
```

Figure 8.20 – Styling for our component

As you can see, the styling we're using here is really not that complicated. So, let's get on with the logic behind how this component works.

As you may remember, I've said we're going to use the state for our price and quantity, so let's initialize our state, as follows:

```
const [quantity, setQuantity] = useState(1);
const [price, setPrice] = useState(0);
```

Figure 8.21 – Initializing state for our component

Now, I was thinking, we could pass the price via a prop; that way, this component is more reusable for future cases. Because this is done via a prop, we should've used a life cycle function, as if we were writing a class component as this is a functional component—and, as we remember, we can use `useEffect` instead of a life cycle function. So, let's import `useEffect` at the same place where `useState` is imported.

Now, let's see how we should write the useEffect function, as follows:

```
useEffect(() => {
    setPrice(props.price);
}, [props.price]);
```

Figure 8.22 – useEffect function used to initialize the price state variable

So, when useEffect gets called, the setPrice function inside of it will get called, which is going to set our price state variable to whatever number the prop is sending. But what's with the [props.price] argument used as the second argument for our useEffect function?

This tells our useEffect function to get called only when the props.price variable gets changed.

Now that we've initialized our price variable, let's change the price based on the quantity. How should we do that? I've written a function called quantityMath that receives a string variable named action that will tell our function whether the quantity should be dropping or rising.

As we all know, when we're shopping online, every item in our basket has a plus and a minus that, whenever pressed, either increments the quantity with one or decrements with one. Based on this, we calculate the total price of that item.

Now, let's take a look at this function, as follows:

```
function quantityMath(action) {
  if (action === "plus") {
    setQuantity(quantity + 1);
    setPrice(price + props.price);
  } else if (action === "minus" && quantity > 1) {
    setQuantity(quantity - 1);
    setPrice(price - props.price);
  }
}
```

Figure 8.23 – quantityMath function used to calculate the final price

Now that we have created this function, let's make sure that when our user presses the buttons, this function is getting called. `TouchableOpacity` is a component used to make other components pressable. So, let's go to one of the `TouchableOpacity` components and change the `onPress` prop to `{() => quantityMath("minus")}`. Of course, we'll use `minus` for the minus icon and `plus` for the plus icon as an argument for our `quantityMath` function. Let's take a look at how that looks in our code, as follows:

```
<TouchableOpacity style={styles.buttons} onPress={() => quantityMath("minus")}>
    <Icon name="minus" family="entypo"/>
</TouchableOpacity>
<Text>{quantity}</Text>
<TouchableOpacity style={styles.buttons} onPress={() => quantityMath("plus")}>
    <Icon name="plus" family="entypo"/>
</TouchableOpacity>
```

Figure 8.24 – Implementing the quantityMath function

Now our component is finished, let's go inside our `App.js` file and test it out. Comment out the previous component and let's import our newly created component. Now, let's add this component to our main function, like this:

```
export default function App() {
    return (
        <View style={styles.container}>
            <CommerceCard itemName="Japanese Painting" price={300} />
        </View>
    );
}
```

Figure 8.25 – Main app function with our CommerceCard component inside of it

Save all files, refresh the app, and you should see our card. Go ahead and start playing with the plus and minus buttons, and you'll see how everything accurately changes based on the quantity you want.

This was pretty cool, right? We now have a cool little component we can use whenever we want to start prototyping for an e-commerce app.

Summary

After learning so much about how React and React Native work, we finally got to the point where more practical challenges are getting tackled head-on. We've started by creating a simple component, where we mostly focused on styling and layout.

That was the easy part, and the first step into our next component was where we saw a different example of creating a layout, and we strengthened our brain muscles so that we can more easily start prototyping components on our own.

Right after this, we got into more serious components, and that was the register form where we learned a new concept called controlled inputs. This was really fascinating as we learned how to actually attack the problem of forms in React Native.

Our next component was even cooler as we used the `useEffect` function to initialize one of our state variables with a prop received by our component. Now, that's some seriously cool stuff, and I hope you got as excited as I did when I first discovered the function.

Now that we've done more practical challenges, it's time to think about how debugging works so that we can make sure we know how to properly find out what's wrong with our component. We'll also learn about some limitations with debugging when it comes to React Native. Let's move on with this cool adventure and get closer to creating our own cross-platform mobile applications.

9
Debugging and Reaching out for Help

We've been through so much already. We've learned how to create different types of components; we've learned about props and state and how each has an important role in our component creation. We've also learned about life cycle functions. We have gained a lot of knowledge so far, but we still haven't got a way of testing our components to see whether they have the behavior we expect.

In this chapter, we're going to learn about debugging and we'll go through the most popular debugging options, such as React DevTools and React Native Debugger. We're also going to learn about some other debugging alternatives so that we can be on the safe side and make sure we use the right tool for the job when needed.

We'll go through interesting concepts such as type checking and linting. We'll also learn about the **Developer** menu and some of the features React Native has for us to quickly find out if our app is having any type of problem.

By the end of this chapter, we should have some knowledge about debugging so we are ready whenever something's not working the way we expect it to work. This will be the last step before creating more complex applications.

The following topics will be covered in this chapter:

- Different ways of debugging
- React Native Debugger
- Where you can reach out for help when you need it

Different ways of debugging

As we all know, developers are human beings and human beings make mistakes. To be completely honest, I feel like software developers make a lot more mistakes than just your average normal human being, so of course, there have to be some ways of solving the bugs that came into existence because of our mistakes.

The process of finding and resolving bugs in computer programming is called *debugging*. There are lots of debugging tactics you can use while solving bugs, so we'll try and get through some of them in this section. Understanding them will surely unlock a new achievement on our React Native journey.

We'll begin this interesting quest of finding out how to make sure there are fewer and fewer errors while we're in the development phase with different formatting tools.

Linting, type checking, and formatting

As developers, we'll mostly want to focus our attention on stuff such as business logic, code patterns, and best practices. You don't usually want to spend time making sure each line is correctly indented or checking what type of argument a certain function needs to receive. To simplify our life and our code writing process, we can make sure all the automation stuff is delegated to our code editor. I'm personally a big **Visual Studio Code** fan, but we've discussed in previous chapters that you may use whatever code editor you want to.

Type checking

The process of verifying and enforcing the constraints of a type is called type checking. This is all to make sure that the possibility of type errors is kept as low as possible. With JavaScript, we don't have to specify what type of information will be stored in a variable and that's all because JavaScript is a loosely typed language. But putting constraints or limitations on our code will make us write more thoughtful code, making us more careful about how we think about the code we're writing.

There are two cool tools when it comes to type checking: **TypeScript** and **Flow**. The main difference between these two is that Flow is just a type checker while TypeScript is a superset of JavaScript, which basically means it will include more next-gen features of JavaScript.

Linting

Linting is the process of executing a program to analyze the potential program syntax errors. The most famous linting plugins for JavaScript are **ESLint**, **JSHint**, and **JSLint**. I personally use ESLint, which now even has an official plugin for TypeScript linting.

You'll see that most people go for ESLint, but that doesn't mean it's the best; you need to figure out what exactly works for you, so try and take a few minutes to google them all. I usually go for the tools with the biggest community just because it's easier to find out how to fix certain errors if they pop up.

Formatting the code

Most of your time as a programmer will be spent reading code, so you'll have to make sure the code you're reading is legible. Let's pretend we want to quickly write a class component; we already know how to do that so maybe we're not even looking at the screen anymore.

Because of that, we're not really paying attention to the way the code looks, but it's irrelevant since we're already good programmers and we know it works. This is how unformatted code looks:

```
export default class Component extends React.Component {render () {return (
<View><Text>Don't write me like this</Text></View>);}}
```

Figure 9.1 – Unformatted class component

I mean… yeah. This doesn't look that good. It works but… where do we even begin understanding what's going on in this big sausage? Now let's see what will happen to our code once we save the file:

```
export default class Component extends React.Component {
  render () {
    return (
      <View>
        <Text>Don't write me like this</Text>
      </View>
    );
  }
}
```

Figure 9.2 – Formatted class component

Phew! It looks 10 times better, right? We can easily follow the code written here. It's a lot easier to read code and understand it when it looks well formatted.

There are multiple different code formatters, but one of the most used ones and also the one that I enjoy using the most is **Prettier**. This is really easy to integrate and configure with your favorite code editor.

By the way, you can even configure your linter to use it for formatting the code so maybe, if you don't really like Prettier, you might actually configure ESLint to do that for you.

In-app Developer menu

There are a bunch of different tools we have access to from inside of our simulator that React Native makes available for us. These are really cool so let's see how to access the in-app Developer menu whenever we're testing our app in the simulator.

The first method of accessing the Developer menu is by shaking the device or selecting **Shake Gesture** inside the **Hardware** menu in the iOS Simulator.

The second method is a keyboard shortcut. For Mac on iOS, the shortcut is *Cmd + D* and for Android it's *Cmd + M*. For Windows, the shortcut is *Ctrl + M* for the Android Simulator. Alternatively, for Android, we can run the following command to open the dev menu:

```
adb shell input keyevent 82
```

Once we've used one of the preceding methods, the following menu will open up:

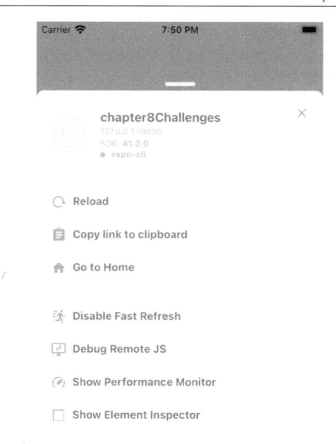

Figure 9.3 – Developer menu

As we can see, there are a bunch of options right here, so let's talk about each one of them. First of all, the ones that are actually interesting to us for debugging purposes are **Debug Remote JS**, **Show Performance Monitor**, and **Show Element Inspector**. Let's begin with the first one.

Debug Remote JS

Clicking this button will open up a new tab in our Chrome browser with the following URL: `http://localhost:8081/debugger-ui`.

Select **Tools** | **Developer Tools** from the Chrome menu to open the Developer Tools. React Native also recommends enabling **Pause on Caught Exceptions** for a better debugging experience. You can do that by going to the **Sources** tab and you'll find this checkbox somewhere on the right, next to the usual buttons used for breakpoints.

Show Performance Monitor

This one is actually pretty cool. Once you click on this button, it'll enable a performance overlay to help you debug performance problems:

Figure 9.4 – Performance overlay

Let's see what we're seeing in the preceding screenshot. We'll begin from left to right, explaining each and every column:

- **RAM** – The amount of RAM your app is using.

- **JSC** – The size of the JavaScript code managed heap. It will only get updated as garbage collection occurs.

- **Views** – The top number is the number of views on the screen and the bottom number is the total number of views in the component. The bottom number is typically larger but usually indicates that you have something that could be improved/refactored.

- **UI** – Main frames per second.

- **JS** – JavaScript frames per second. This is the JavaScript thread where all the business logic lives. If the JavaScript thread is unresponsive for a frame, it will be considered a dropped frame.

Show Element Inspector

Here it is! The last option in our Developer menu. Let's click it and see what happens. Our screen has kind of changed:

Figure 9.5 – Element Inspector once we enable it

Now that we've clicked it, we can see that it asks us to tap on something so it can inspect it. At the same time, we also can see there are four different tabs down there called **Inspect**, **Perf**, **Network**, and **Touchables**.

These can all be used just like you'd use the Chrome Developer Tools, but with more limitations, so you'd probably prefer using the Developer Tools. Let's at least tap on an element and see how it appears once we click it:

Figure 9.6 – Our Element Inspector once we have clicked the commerce card

Once we click the commerce card, we can see it has a blueish overlay on top of it with a green border. That green border represents the padding. But let's focus our attention on the upper part of the screen where our inspector has now moved.

In the upper part of the inspector, we can see the component tree, which basically tells us what component exactly we have clicked. So, we've clicked a `View` component inside a `Block` component, which lies in a `Context.Consumer` component. I guess we can read even further than that and see that this is all part of the `CommerceCard` we created in the previous chapter.

Underneath the component tree, we have the styling applied on the View we've clicked. Toward its right, we have information about the *size*, *padding*, and *margin*.

The best way of actually learning how to use all these internal tools that are provided to us by the React and Expo team is to actually play around with them. You probably won't use these as much as the following tool, but I'm pretty sure you'll want to experiment with them. The following tool is one of the most commonly used for debugging.

React Native Debugger

React Native Debugger includes almost all the tools necessary for debugging a React Native application. That's why I totally recommend using this one as it has everything you need inside it.

This is basically a standalone app based on the official **Remote Debugger** but with more features implemented. It also includes **React Inspector**, **Redux DevTools**, and **Apollo Client DevTools**. We're not really interested in Redux and Apollo right now, but you'll most probably stumble upon *Redux* as it's one of the most used libraries for state management.

You can install React Native Debugger on macOS via the following command:

```
brew install --cask react-native-debugger
```

If this command doesn't work, you should make sure you have **Homebrew** installed. Homebrew is a module manager and you'll for sure keep on using it with different programming tools. To install Homebrew, visit `https://brew.sh`.

To install React Native Debugger on Windows, we have to go to the following URL: `https://github.com/jhen0409/react-native-debugger/releases`. Download the `.exe` file and open it up.

Now that the software is opened up, press *Ctrl + T* on Windows or *Cmd + T* if you're on a Mac. This will open up a new window where you'll be prompted to specify the port. Write 19000 there and click **Confirm**:

Figure 9.7 – Window opened for changing the port

Now we can run our project with expo start or expo r -c. After that, open up the **Developer** menu and select **Debug Remote JS**. The debugger should automatically connect now.

Now you should be able to see the element tree as well as the props state and children of whatever element you've selected. On the right, you'll see the Chrome console:

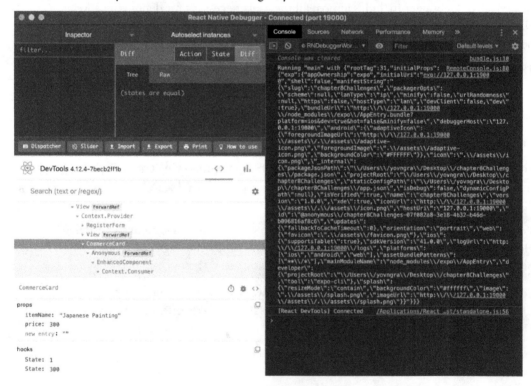

Figure 9.8 – React Native Debugger connected to our simulator

If you right-click anywhere in the React Native Debugger, you'll see we have some cool little shortcuts that we can use to reload our app, enable the element inspector or network inspector, while also clearing our **AsyncStorage** content.

We can even use this one to inspect our network traffic so right-click anywhere and select **Enable Network Inspect**. This will enable the **Network** tab and allow us to inspect `fetch` or `XMLHttpRequest` requests. Because there are some limitations to inspecting networks using React Native Debugger, you might want to look for some alternatives. All of them require a proxy but here are some alternatives you might want to look into: *Charles Proxy*, *mitmproxy*, and *Fiddler*.

As we know, React Native Debugger has React DevTools implemented inside of it, so maybe you don't want to mess with all the tools at once and you'd really love seeing the component tree with some properties.

Even though we've installed React Native Debugger, I'd really recommend at least keeping in mind that we can also use each tool included in it but separately.

React DevTools

This tool is really great for checking out the component tree and each component's props and state. First, if we want to install it, we need to do it via `npm` with the following command:

```
npm install -g react-devtools
```

This will install React DevTools globally on your computer but you might want to just install it as a project dependency. If that's the case, you can do that via the following command:

```
npm install --dev react-devtools
```

Now that we have React DevTools installed on our computer or project, let's start up our project with the usual `expo start` command. After we've opened up our project, let's open a new terminal window and run the following command:

```
react-devtools
```

This will open up a new window. Now we need to open the Developer menu inside our simulator and click **Debug Remote JS**. It's the same process as before but we don't need to set up the port with React DevTools because it will automatically connect to our project. We can see how the app looks by looking at the following screenshot:

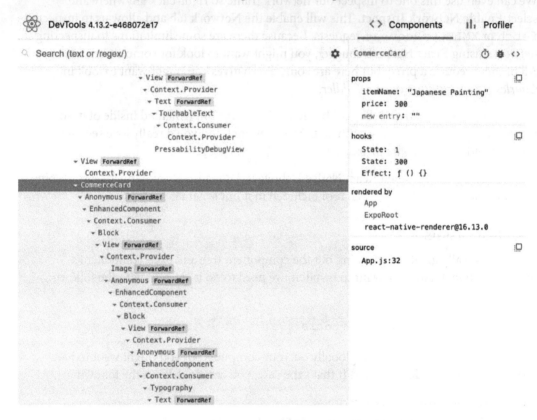

Figure 9.9 – DevTools standalone app for Debug Remote JS

As far as we can see, this is identical to our bottom-left window in React Native Debugger. I'll mostly use this because it makes it easier for me to check out my component but as the app gets bigger, you'll probably catch me switching to React Native Debugger.

All in all, this is a great tool to have under your belt and I highly recommend playing around with it if you don't really have too much experience with Chrome's Developer Tools as these tools are really similar to what a web developer is used to.

Now that we have found out about some tools used for debugging React Native applications, let's see what else can we do if an issue can't really be debugged with these tools. Or, maybe some of you might even see these tools as too much work so let's check out other solutions to some of the problems we might encounter.

Where you can reach out for help when you need it

I know for a fact that almost all programmers get stuck at some point while developing a product or a feature for an already existing product. So, we should know what to do when an error pops up.

Sometimes, you can tell exactly what's wrong just by the **stacktrace**, but other errors might be a bit more difficult to understand at first glance. The stacktrace is the big red error message that pops up on your simulator whenever there's a problem with your code.

First of all, I think you should know that because we're on React Native and the community is so big, almost all error messages can be searched on Google. There's always going to be someone out there with a solution for your error.

Another good solution would be to isolate the code that's throwing the error. You can do that by discovering which line exactly threw the error and then commenting out that section. By isolating the code, we can start experimenting with that part alone and by trial and error, we can get to a working solution.

A really good habit that you should start developing is the use of `console.log`. You can use this to discover how your code works. For example, by using it before and after we do something with a state variable, we can see how exactly the variable changes by constantly following it inside our code. The only issue with using `console.log` instead of breakpoints with a debugger is the fact that when we have any type of `async` code, we might not realize that some code is changing at different points that might be out of our control.

If you can simplify your code as much as possible, you might be able to track down errors much more easily than you normally would be able to. Because of that, you'll see that some repositories on GitHub ask for a *minimal reproducible demo* in their bug reports. This allows them to see that you correctly identified the issue and isolated it. So, if the app you're working on is a bit too large and complex, extract the functionality and try to identify the specific error.

Of course, you might also get into some production errors. Some errors and bugs might show up only in production mode. So, it's a good thing to test your app once in a while in production mode by running the following command:

```
expo start --no-dev --minify
```

The best first step in understanding a production error is to reproduce it locally. After that, just isolate the issue and find a good solution for it.

The Expo team recommends an automated error logging system such as Sentry. This tool will help you in tracking, identifying, and resolving JavaScript errors in your production app. It even provides you with *sourcemaps* so you will have stacktraces of your errors. This tool is free for up to 5,000 events a month.

Let's think of what we would do if our production app was crashing. What could be the cause of that? This is one of those really frustrating scenarios at first glance. The truth is this is pretty easy to understand and solve.

The first thing you should do is to access the *native device logs*. You can do that by following the instructions for whatever platform you're using. We will see how to check logs on each of these platforms in the next sections.

Logs for iPhone/iPad

Follow these steps to check your iPhone logs:

1. Open up your terminal and use the following command:

    ```
    brew install --HEAD libimobiledevice -g
    ```

2. Now that this package has been installed, plug in your iOS device and run the following command:

    ```
    idevicepair pair
    ```

3. Click the **Accept** button on your device.

4. Go back to the computer and run the following command:

    ```
    idevicesyslog
    ```

Congratulations! Now you can check out your iPhone logs.

Logs for Android

Make sure your Android SDK is installed. Make sure that USB debugging is enabled for your device. If it isn't enabled, you should be able to find out how to do that at https://developer.android.com. The information you're looking for should be under **User Guide | Build and run your app | Run apps on a hardware device**.

Now plug in your device and run `adb logcat` inside the terminal.

Congratulations! Now you're able to check out your Android logs.

This is great! We've found out how to check our logs and this should point you in the right direction in your bug-solving adventure. Search the logs with the words "fatal exception," and this should quickly point you to errors. Now reproduce the errors! By reproducing them, we're making sure that our assumption of how exactly they behave will be proven.

Okay, but what if my app only crashes on a specific or older device? This has a 90% likelihood of being an indication of performance issues. The best thing you could do in this situation is to run your app through a profiler to see what exactly is killing your app. Hmm, do we know a good profiler? Yes, React DevTools or React Native Debugger both have a profiler included. I'd honestly recommend you read `https://reactnative.dev/docs/profiling` because there's a ton of information about how exactly to identify which processes take high amounts of memory and could potentially kill your app.

Still can't figure out what's wrong with your app?

This is the perfect moment to seriously consider taking a break. I know it might sound weird but a 10-minute break is a lifesaver in certain situations. I sometimes even sleep on the issue until the next day and once I open up Visual Studio Code, the solution kind of comes to me.

You could also just try a Google search again. The best places to find solutions are the **Issues** section on GitHub, Stack Overflow, Reddit, and Expo forums.

Summary

This chapter wasn't as expansive as the others but I hope you were able to find all the information necessary for a jumpstart into how exactly to solve all the issues you might encounter while developing with React Native and Galio.

We've been through certain tools to prevent mistakes in our code writing. I highly recommend going over all of them and doing more research because, as we all know, knowledge is power. The more tools you learn about, the better you'll feel once you find the perfect match for you.

After going through those tools, we learned about React Native's built-in tools for debugging and profiling. We learned how to use the features found in the Developer menu and I hope you understand the fact that even though the information presented here is brief, the most important thing is for you to experiment with all these tools.

We've also learned about React DevTools and React Native Debugger. Now that we know how to install and start up these tools, it should be fairly easy to experiment with our apps to understand more about how exactly React Native works.

We also learned some ways and tactics for finding out where an error's coming from. I really do hope I've done a good job of explaining these topics as they usually come packed with your programming experience. Even though I understand debugging is not the most exciting experience, it is part of the job and it's really cool to learn about it when you get to the point where you actually need it.

Now let's move forward because it's time for some practical challenges! We'll start by building the Onboarding screen for the Stopwatch app that we're going to create further down the road in this book. I really hope you're ready for some neat tricks as the Onboarding screen is going to teach us a lot about `FlatList` and how to use a reference to control a component via another component. Now, get ready, set, go!

10
Building an Onboarding Screen

We've been through so much that we can say it's finally time to start building more than just components. We'll begin by creating a really important part of any application and that's the onboarding screen.

We'll go over what exactly an onboarding screen is and its purpose in an app. We'll understand that there are many types of onboarding screens, but we'll focus on creating just one of those types.

By learning how to create this type of screen, we'll learn lots of cool new concepts we haven't been exposed to until now. These new concepts will be helpful for you in the future for building lots of different types of screens. By learning a lot of new things, we can surpass our creative limitations and be more prepared for future challenges.

We'll learn about animation and how to create a cool animation for our screens. This will open the door to creating a smoother user experience for our customers. We'll understand what interpolation and extrapolation mean and how we can use them to build animations.

We'll also go more in-depth regarding Hooks and how to use `useRef`. And yes, we will also learn about a new surprise Hook that is going to help us find the size of our screens even faster than before.

Last but not least, we'll learn how to use a cool component that's more performant than any tool we've used until now. This cool component is called `FlatList` and it's going to help us create a cool onboarding experience for us and our users.

By the end of this chapter, we'll have a great onboarding screen that we'll use as the main opening screen for the next chapter's app project.

This chapter will cover the following topics:

- What's an onboarding screen and where can we use it?
- Creating a new project
- Paginator
- Adding the automatic scrolling functionality

Technical requirements

You can check out this chapter's code by going to GitHub at `https://github.com/PacktPublishing/Lightning-Fast-Mobile-App-Development-with-Galio`. You'll find a folder called `Chapter 10` that contains all the code we've written in this chapter. To use that project, please follow the instructions in the `README.md` file.

What's an onboarding screen and where can we use it?

We should start this chapter by understanding what exactly an onboarding screen is. An onboarding screen is like a short introduction to your app before it's used. It's the first screen that welcomes the user.

The onboarding of your app should have a specific goal when it comes to welcoming users. You'll have to make sure that your onboarding will help users understand how they're supposed to be using the app, while also exciting them about the features they'll be able to use.

If you've made sure that the onboarding screen is going to be a great experience for your users, then you can expect more engagement from them in the first few days of them using the app. High engagement means happy users, which means your app is creating a really good user experience.

The onboarding screen should only appear to first-time users. We all know how annoying it would be to redo a tutorial in a game. Even though it should only take about 30 seconds to 1 minute to go through it, this could still make a returning user annoyed with the experience.

I'd recommend reading more about onboardings by going to Google's Material website, where they recommend different design ideas and guidelines for creating a good onboarding screen for Android phones: `https://material.io/design/communication/onboarding.html`. Of course, most of the same rules apply to iOS as well.

Now that we've figured out what an onboarding screen is, as well as where and when to use one, it's time for us to figure out what exactly this onboarding screen would look like for our app. Also, this is a good time for us to talk about what type of app this will be.

The next chapter is going to focus solely on the logic part of our app, while this chapter will focus on creating the onboarding we need for our app. The app in question will be a stopwatch/timer app. Knowing this, I decided to have an onboarding experience where the user learns about the utility of the app.

Let's take a look at what our onboarding screen for this app will look like:

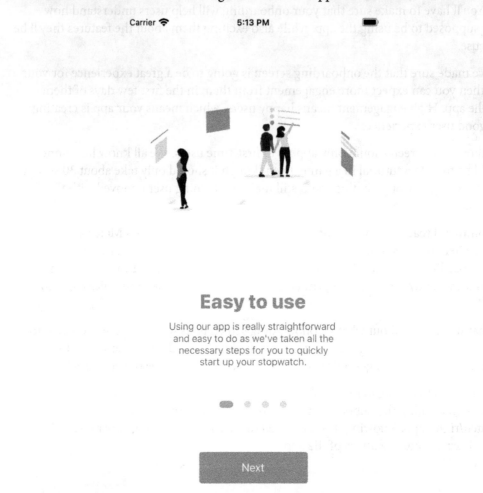

Figure 10.1 – The onboarding screen we're going to create

This looks pretty cool, right? It will be formed of four different screens, each of which will have a different image, title, and description. We'll have the ability to swipe left/right to advance and read all the screens while also pressing the **Next** button, which is going to swipe for us. The small four dots behind the text are going to be animated to show us which screen we are on.

I honestly love the way it looks, and I can't wait to finish this onboarding screen so that we can start creating the full app. By the end of this chapter, you'll be ready to start creating onboarding screens.

Now, let's start working on our screen.

Creating a new project

Now that we know what our project will look like and why we need this screen, it's time to start working on it.

Before coding, let's gather the images that we're going to use for this project. I've used the images from `https://undraw.co`, which provides open source `.svg` and `.png` images. I've downloaded four different `.png` images and placed them in the `assets/onboarding` folder. The `onboarding` folder is a new folder I've created in the `assets` folder specifically for this screen.

Let's start by opening the terminal window and moving to the directory that you usually use for your projects. Now, let's write our usual command and get right into it:

```
expo init chapter10
```

Now that we have a new project, let's install Galio. We can do this by using the following terminal command:

```
npm i galio-framework
```

Now that we've set everything up, let's open our project and start coding.

First, we'll begin by doing the same old trick we've always done, and that is going into our `App.js` file and deleting everything inside `View` in our `App()` function. I've left out the `<StatusBar />` component just for the sake of it as you may wish to style or hide it.

Now, in our main folder, let's create a folder called `components`. This is where we'll place every component we're going to create for this beautiful app. By the end of this chapter, you'll have three files in this folder.

Now that our folder has been created, create a new file inside it called `Onboarding.js`. This will serve as our main screen. Now, let's quickly create and export this component so that we can import it into our main `App.js` file.

We've already done this a lot of times but this time we'll be using `SafeAreaView` instead of a `Block` or `View` component as the parent component for our screen. We're using this one because we want to make sure everything is taken care of in case the users of our app have a phone with a notch:

```jsx
import React from 'react';
import { StyleSheet, SafeAreaView } from 'react-native';
import { Text } from 'galio-framework';

export default function Onboarding() {
  return (
    <SafeAreaView style={styles.container}>
      <Text>Onboarding</Text>
    </SafeAreaView>
  );
}

const styles = StyleSheet.create({
  container: {
    flex: 1,
    alignItems: 'center',
    justifyContent: 'center',
  }
});
```

Figure 10.2 – Onboarding component ready to be imported

Now that we've finished coding this function, let's go ahead and import it into `App.js`. Once imported, we can use it inside our main function. Now, the main function should look something like this:

```jsx
export default function App() {
  return (
    <View style={styles.container}>
      <Onboarding />
      <StatusBar style="auto" />
    </View>
  );
}
```

Figure 10.3 – The main App function with Onboarding imported

Now, we're ready to continue developing our app. Because we've imported the Onboarding component inside our App function, we can see whatever we're going to modify and add our Onboarding component every time we save the file.

Let's get back to Onboarding.js and think of how we should start working on our onboarding screen.

By looking at *Figure 10.1*, we know that there are three main identifiable parts to this screen. Remember when we discussed that we need to always look at screens in terms of bigger containers to understand how to create the layout before we start coding? Well, this is the same thing, so let's split our screen into those three main parts:

- **OnboardingItem**: This area is the upper part of the screen, before the four dots. It's going to include a picture, a title, and a description. This area needs to be set as one big area because it's going to change with every swipe.

- **Pagination**: The four small dots display where we're situated in this big slider. This way, we always know how much more there is to read until we get to the last slide.

- **Next button**: This button is going to move the slides without the need to swipe, while also being the last thing we need to press to move on from the onboarding screen to the home screen.

Knowing that there are three parts should make things easier to sketch. First of all, we should start with the main area, which is also the biggest one. Let's go to our components folder and create a new file called OnboardingItem.js.

OnboardingItem

As we mentioned previously, this component should render the top of our screen, which contains a picture, some text, and a description. Let's do just that. Go ahead and import the Block and Text components from 'galio-framework' and Image and StyleSheet from 'react-native'.

Once we've imported everything we need, it's time to start building our component. We'll begin by having a main <Block /> component, which will be hosted inside our <Image />, and another <Block /> component. The second <Block /> component will have two <Text /> components as children – one for the title and one for the description. This second <Block /> component will be used to split the screen into the main area, which is going to take up some more space because it's going to be an image, and the secondary area, which should be smaller because it only contains text:

```javascript
import React from 'react'
import { StyleSheet, Image, useWindowDimensions } from 'react-native';
import { Block, Text } from 'galio-framework';

export default function OnboardingItem({item}) {
    const { width } = useWindowDimensions(); //get the width of the screen with this hook

    return (
        <Block style={[styles.container, {width}]}>
            <Image source={item.image} style={[styles.image, {width, resizeMode: 'contain'}]}/>
            <Block flex={0.3}>
                <Text style={styles.title}>{item.title}</Text>
                <Text style={styles.description}>{item.description}</Text>
            </Block>
        </Block>
    );
}
```

Figure 10.4 – The OnboardingItem component

As you can see, there's a small surprise here. First of all, I've imported useWindowDimensions. This is a hook function that returns the width or height of the screen. This is an easy way to make sure that the width of your components is equal to the width of your screen.

I've also imported a prop called item. Because this is an onboarding screen, we'll have at least four screens with a different type of text or image. We'll pass an object, along with all the necessary information, through this prop called item. This way, we can make sure everything will go to the exact spot we want it to go, as well as that we won't have to keep wasting time writing props for every single part of our component.

All the styles have been applied already, but we haven't discussed them much yet. So, let's look at these styles for a second:

```
const styles = StyleSheet.create({
    container: {
        flex: 1,
        alignItems: 'center',
        justifyContent: 'center',
    },
    image: {
        flex: 0.7,
        justifyContent: 'center'
    },
    title: {
        fontWeight: '800',
        fontSize: 28,
        marginBottom: 10,
        color: '#c9a0dc',
        textAlign: 'center'
    },
    description: {
        fontWeight: '300',
        color: 'gray',
        textAlign: 'center',
        paddingHorizontal: 64
    }
});
```

Figure 10.5 – Styles for our OnboardingItem component

As we can see, the container has a property of `flex: 1`, which tells our main `<Block />` component to take up as much space as it can. We gave our `image` a `flex: 0.7` property because we want it to occupy 70% of the space, while `title` and `description` only need to occupy 30% of the space. The other styles are just the usual text styles where we set `fontSize` and `fontWeight`.

> **Tip of the Day**
>
> I cannot stress enough the importance of looking at the pictures to understand the code. I believe you should look at the picture first, try to make all the connections inside your brain, and then see if you got *lucky*. I wouldn't call it luck, though; I'd call it more of an educated guess.

Now that our `OnboardingItem` component has been created, we're ready to import it into our `Onboarding.js` file, which is where our main screen resides. We all know how to import a component but just to make sure, the line we must write for this looks as follows:

```
import OnboardingItem from './OnboardingItem';
```

Now that we've done this, we can start using `FlatList` to render all our items on the screen.

FlatList

As we mentioned earlier, we don't want to repeat ourselves, so we won't write the same code four times inside our main onboarding component. The first thing that may come to your mind is to use the .map() function. This is a good guess, but React Native has a component that is usually used because of its better performance. It also has some props built into it that are lifesavers in situations like this. This component is called FlatList.

To use this component, we need to prove it with an array of elements that must be mapped to the component we've been creating. Earlier, I mentioned that our OnboardingItem component will accept a prop called item. This prop is going to be an object inside this array.

If we look at *Figure 10.4*, we can figure out what our object should look like from the way we've used it inside our component. We know that it needs to have a title, description, and image, but it also needs an id.

Let's create a new file called slides.js in our root (main) folder. This is where the array containing all the objects that provide the information our onboarding needs will reside, as shown in the following screenshot:

```
export default [
    {
        id: '1',
        title: 'Easy to use',
        description: 'Using our app is really straight forward and easy to do as
        image: require('./assets/onboarding/show-around.png')
    },
    {
        id: '2',
        title: 'Run, Forest, run!',
        description: 'You can easily check your time for each lap',
        image: require('./assets/onboarding/progress.png')
    },
    {
        id: '3',
        title: 'Timer',
        description: 'With a great timer at your disposition you\'ll always know
        image: require('./assets/onboarding/bytheroad.png')
    },
    {
        id: '4',
        title: 'Ready to go!',
        description: 'Thanks for using our app! I hope you are going to love it!',
        image: require('./assets/onboarding/completion.png')
    }
];
```

Figure 10.6 – The array with all the information needed by the FlatList component

Don't forget that you can have any type of information you'd like. The titles, descriptions, or images don't have to be identical to mine. Remember that when we started creating our app, I told you to download some images and place them inside the `./assets/onboarding` folder. These are the images I've chosen, and I imported them using the `require` keyword.

The `require` keyword is used just like `import` in that it tells the JavaScript engine that it needs the file to be located at the specified destination.

Now that we have the array with data for `FlatList` ready it's time to go back to our `Onboarding.js` file and import this new file just like this:

```
import slides from '../slides';
```

Now, let's make sure we have the rest of the imports ready for when we need them as we need some more components. First of all, we'll remove the `Text` import and we'll import the `Block` and `Button` components from `'galio-framework'`. Secondly, we'll add `FlatList` to the list of imported components:

```
import React from 'react';
import { StyleSheet, SafeAreaView, FlatList } from 'react-native';
import { Block, Button } from 'galio-framework';

// components
import OnboardingItem from './OnboardingItem';

import slides from '../slides'; // data
```

Figure 10.7 – New imports added to the Onboarding.js file

Now that everything has been imported, we're ready to start developing the screen. We'll delete our `<Text />` component from within the `<SafeAreaView />` component and use a `<Block />` component with the `flex={3}` prop instead.

Inside of this `<Block />` component, our `<FlatList />` component will start its own life. Let's take a look at how I've implemented this component before explaining how it works:

```
export default function Onboarding() {
  return (
    <SafeAreaView style={styles.container}>
      <Block flex={3}>
        <FlatList
          data={slides}
          renderItem={({item}) => <OnboardingItem item={item} />}
          horizontal
          showsHorizontalScrollIndicator={false}
          pagingEnabled
          bounces={false}
          keyExtractor={(item) => item.id}
        />
      </Block>
    </SafeAreaView>
  );
}
```

Figure 10.8 – FlatList implemented in our onboarding screen

As you can see, implementing this seems pretty straightforward. By the way, if you were to copy this code into your editor right now (assuming you've been following along with everything else as well) and try to check out your app on the simulator, you would see a fully working onboarding screen. Yes, it doesn't look as good as what we showcased at the beginning of this chapter, but it's working – we can slide left and right and check all the information we've been writing in our `slides.js` file.

Now, let's look at how this component works:

1. First, we'll begin with the `data` prop. This is where we're offering our `FlatList` component the array it needs to start rendering each element.

2. Then, we have the `renderItem` prop, which is where we use a function to render the item we need. In this case, we need multiple instances of `<OnboardingItem />`.

    Remember when I said we're going to pass this component the prop called `item`? This is because we only need to pass an object from the array. Our `FlatList` component will pass each object to a different `<OnboardingItem />` component. Once we've done this, we can capture that object and use it in any way we see fit.

3. The `keyExtractor` prop is used to extract a unique key for a specific item at the respective index.

    These keys are used for caching so that React can keep track of each item individually and only rerender the one that must be rerendered. Do you remember how we were using the `key` prop when we were rendering items with the `.map()` function? This is the same but all the work is done by this prop. This is why we needed an `id` key inside our objects for the `slides` array.

4. The rest of the props are used mainly for layout purposes. I'd strongly encourage you just play around with those props by turning them on and off. For example, the `horizontal` prop makes our list, well, horizontal.

Now that we've successfully built our list of elements, which is the first step of creating a great onboarding experience, let's start building the paginator.

Paginator

The paginator is displayed on the screen by those four small dots. Its main purpose is to show the user which slide they're currently looking at while also displaying a sense of progress. This small component is not that hard to implement, but the features we're going to use to make sure this thing is working correctly are new to us.

One of the most important objects we're going to work with for this component is the `Animated` object. This is needed because we're looking at animating the width and the opacity of our dots. This is also important because we want to make sure that the animation happens at the right moment. The right moment is, of course, while the user interacts with `FlatList`. If your finger moves from right to left, we want the animation to also move at the same pace as your finger.

We're also going to use a cool new Hook called `useRef`. This Hook is used when we need a mutable object that persists for the entire lifetime of your component. `useRef` will not cause your component to rerender when its value is changed because it's not a state variable, but it's a really good way to make sure that you'll get the same `ref` object on every render.

So, let's get started on this cool little component, which I'm sure you're going to find helpful and reusable for future applications. We'll start in Onboarding.js. Let's begin by importing useState and useRef from 'react'. We'll also import Animated from 'react-native'. After importing all of this, we're ready to move on:

```
import React, {useState, useRef} from 'react';
import { StyleSheet, SafeAreaView, FlatList, Animated } from 'react-native';
import { Block, Button } from 'galio-framework';
```

Figure 10.9 – Fresh new imports for our Onboarding component

Now, let's start implementing everything we need for our Paginator component. We'll begin by creating two new ref objects that we're going to implement in our FlatList component:

```
const scrollX = useRef(new Animated.Value(0)).current;
const viewConfig = useRef({ viewAreaCoveragePercentThreshold: 50 }).current;
```

Figure 10.10 – Our newly created refs

Let's explain how this works. First of all, we'll begin with scrollX. This variable has a lot going on, so let's start from the beginning. We're creating a new ref object using the useRef Hook and we're initializing this new variable with Animated.Value(0).

Animated.Value creates a value that can be animated. If we were to initialize this variable with just a number, such as 0, React Native wouldn't know how to handle it when it comes to animations.

The useRef Hook returns an object that looks like this:

```
{ current: … }
```

To access the value stored in current, we must write scrollX.current. A workaround for this is to let JavaScript know we want to access that value by chaining .current to our useRef Hook.

The viewConfig variable works just as you'd expect it to work. Here, we must create a new ref object and initialize it with the object shown in *Figure 10.10*, { viewAreaCoveragePercentThreshold: 50 }. Now, let's use these two new variables with our FlatList component:

```
return (
  <SafeAreaView style={styles.container}>
    <Block flex={3}>
      <FlatList
        data={slides}
        renderItem={({item}) => <OnboardingItem item={item} />}
        horizontal
        showsHorizontalScrollIndicator={false}
        pagingEnabled
        bounces={false}
        keyExtractor={(item) => item.id}
        viewabilityConfig={viewConfig}
        onScroll={Animated.event(
          [{ nativeEvent: { contentOffset:  { x: scrollX }}}],
          {
            useNativeDriver: false
          }
        )}
      />
    </Block>
  </SafeAreaView>
);
```

Figure 10.11 – Implementing our new variables inside FlatList

Things might look a bit complicated right now, but it's a lot easier than it looks. With that, we've added two new props to our <FlatList /> component called onScroll and viewabilityConfig.

The viewabilityConfig prop is here to support our pagingEnabled prop, which is telling our list of components to move to the next or previous slide based on how far the user has swiped. By setting the viewAreaCoveragePercentThreshold value of viewabilityConfig to 50, we're telling our component to only go to the next slide if the user has already swiped more than or equal to 50% of our current slide.

The onScroll prop is firing a function every time the user scrolls through the slides of our onboarding screen. You might be wondering what Animated.event: does? It maps an animated value to an event value. I agree, the function looks pretty chaotic, but it's pretty easy to understand if we learn how to read it. So, we're mapping our scrollX value to the nativeEvent.contentOffset.x event value. This event value is usually passed to the onScroll callback, so remember that you might see or use this more often than you think.

The `Animated.event` function accepts two arguments. The first argument is the array of values that we're going to want to map to our **Animated** values. This `event()` function does this by calling the `setValue()` function on the mapped outputs; in this case, on `scrollX`. The second argument is a `configuration` object, which we're using to tell React Native that we don't want to use the native drivers for our animation.

You might assume that by using the native drivers, we might have better performance and this is correct. The reason why we don't want to use the native drivers in this specific use case is because we're going to animate the width of our dots and right now, React Native can't use the native driver to animate width or layout properties in general.

Now that we know why we need `scrollX` and `viewConfig`, we should start building our new component. Create a new file inside the `components/` folder called `Paginator.js`. Now that we've created a new file, we should start building our functional component.

We'll start by importing everything necessary from `'react-native'`; that is, `StyleSheet`, `View`, `Animated`, and `useWindowDimensions`. The next step is building our function:

```
export default function Paginator({ data, scrollX }) {
    const { width } = useWindowDimensions();
    return (
        <View style={{ flexDirection: 'row', height: 64 }}>
            {data.map((_, i) => {
                const inputRange = [(i - 1) * width, i * width, (i+1) * width];

                const dotWidth = scrollX.interpolate({
                    inputRange,
                    outputRange: [10, 20, 10],
                    extrapolate: 'clamp'
                });

                return <Animated.View style={[styles.dot, {width: dotWidth}]} key={i.toString()} />
            })}
        </View>
    )
}
```

Figure 10.12 – Paginator component almost completed

There's a bunch of new stuff here, so let's start explaining everything from top to bottom.

This component, which we've called `Paginator`, accepts two props called `data` and `scrollX`. `data` is the array of objects we passed to `FlatList`, while `scrollX` is the `Animated` value we defined in our `Onboarding.js` file (our parent component).

We've already discussed that the `useWindowDimensions()` Hook returns the `width` and `height` properties of our screen, so that should be easy to understand.

We've given the styles of flexDirection: 'row' and a height of 64px to our <View /> component, which contains the *soul* of our component. We've done this to make sure the dots we'll be creating are going to be sitting nicely in a row.

After that, we used the .map() function to map over the array. As you can see, the map() function accepts a callback function that takes in two arguments. The first one, _, will be our element, while the second one, i, is going to give us the index of that element.

So, for each element in our array, we're creating a dot. How do we do that? Let's jump straight to our return statement to find out. Here, we're returning a <View /> component with styles.dot applied to it. The reason why we're calling it <Animated.View /> is because we want to animate this component. But before we start to animate it, this could just be a normal <View /> component:

```
const styles = StyleSheet.create({
    dot: {
        height: 10,
        borderRadius: 5,
        backgroundColor: '#d4b3e3',
        marginHorizontal: 8,
    }
});
```

Figure 10.13 – Styles for our dots

These are the styles we were using to create the dots. As you can see, there is no width, which is because we want to animate the width of the dot. However, if we were to never animate it, we could've just gone right ahead and given it a width of 10px.

So, let's get back to how we animate the width of our dots. As you can see, we have a variable called inputRange that is an array of values based on the width of the screen and our dot's index. As we know, a slide occupies the full width of the screen. Knowing that, we can understand that a slide has changed when contentOffset.x is equal to the width of the screen. It's called contentOffset because it gives us the offset between two elements. It starts at 0 when the first slide is on the screen. Once that slide moves out of the screen and the next slide comes in, the difference between the last slide and the next one is equal to the width of the screen. Understanding how contentOffset works enables us to think of a way to start creating animations.

What exactly makes an animation? I feel like this is a great place where we can define how exactly an animation works. Let's imagine we have a box on the screen and whenever someone presses a button, we want that box to appear. Of course, it can suddenly appear on the screen, but that doesn't look that good. This is where animations come in. Instead of suddenly appearing on the screen, what if we had a smoother transition? What if the box transitioned into existence over some time? That would look more like an animation, right?

This is the same concept we're applying here: we want the movement of our dots to be completely in sync with the movement of our slides. So, we need our dot's width to grow at the same time as we move our finger on the screen because this creates a smoother experience for our users.

Keeping that in mind, we've mapped the animated value of `scrollX` to our `nativeEvent.contentOffset.x` event value. Now, we can access the exact amount of change between two elements in our horizontal list via `scrollx`. Based on that amount, we need to change the width.

But there's a *problem*: our dot's `height` is `10px`, so if we want our dot to be, well, a dot, then we'd need `width` to be `10px` as well. The problem is that our `scrollX` will go way beyond `10px` because our screen's width is bigger, so how can we let React Native know that we want our current dot to have a bigger width and the rest of the dots to have a width of `10px`? With **interpolation**.

Interpolation

So, let's recap for a second. We want the dot corresponding to the slide we're viewing at this moment to have a bigger width (let's say, `20px`) than the ones that are corresponding to the slides that are out of our view. The only way we can do this is with interpolation.

Interpolation is the way we're estimating the output of a function based on the input we've provided.

Let's assume we have a function where all we know about it is that $f(0) = 0$ and $f(10) = 20$. Can you guess what $f(5)$ will be equal to?

x	0	5	10
f(x)	0	?	20

Based on this table, we could suggest `10` as the answer to our question because 5 is between `0` and `10` and we know the answer should be between `0` and `20`. This intuitive approach is what interpolation does.

So, now that we know how our values need to behave, we can take a look at the interpolation function for our dot's width:

```
const dotWidth = scrollX.interpolate({
    inputRange,
    outputRange: [10, 20, 10],
    extrapolate: 'clamp'
});
```

Figure 10.14 – The interpolate function

So, we want this function to return a value between 10 and 20 based on the user's current position. Our inputRange variable, as we mentioned earlier, is defined by the index of that specific slide and the width of the screen. The first value in our inputRange variable is represented by the previous slide, the second value is represented by the current slide, and the third value is represented by the next slide. Based on that input, we've created an outputRange where we know that the previous slide's dot should have a width of 10px, the current slide dot's width should be 20px, and the next slide's dot width should be 10px.

Guessing which value should be returned based on inputRange is React Native's job, but what we're really interested in is the value itself. Now, we can go to our <Animated. View /> component and have the width of each dot equal to dotWidth, which is the value that's given to us by interpolation. Now, the width will change at the same time as the user swipes their finger.

Extrapolation

We also have another cool little thing here called extrapolate. So, we know our inputRange is only taking the previous, current, and next slide into consideration, but what about the fourth one? Because we haven't specified any value for the fourth one, React Native can start guessing what the width should be for itself.

If we run the code without a little help from extrapolation, we might see some weird results. I encourage you to delete the extrapolate line and see what happens for yourself.

We can solve those weird results by adding extrapolate to our interpolate function. This will tell React Native *what should happen outside of the ranges we've provided* and what kind of pattern the outside values should follow. This is great when we have no idea what the boundaries of our range are. In this case, the solution will be to **clamp** your ranges. This means that whatever comes before or after that range, we'll keep the last given value.

By using `extrapolate: 'clamp'`, you're going to clamp the range from both sides, but if a specific case requires it, you can always clamp only the side of the range that you need. This means you could clamp to the left of the range or the right as well.

> **Tip**
>
> The default mode for extrapolation is `extend`, which is React Native guessing the values of our range.

Great! Now that we've explained how to interpolate and extrapolate, we've understood how (and based on what) the `dotWidth` variable changes. Because this is all done with a `scrollX` animated value, we've placed the `dotWidth` variable inside a `<Animated.View />`. Now, our width changes based on our scrolling behavior.

What's left? Well, I feel like it'd be cool to see the opacity changing as well. The current dot should have an opacity equal to `1`, while the other dots should have an opacity of `0.4`. Based on this information, try to do this by yourself.

If you couldn't do it, worry not! This is a lot easier than it may seem. Let me show you!

```
const opacity = scrollX.interpolate({
    inputRange,
    outputRange: [0.4, 1, 0.4],
    extrapolate: 'clamp'
});

return <Animated.View style={[styles.dot, {width: dotWidth, opacity}]} key={i.toString()} />
```

Figure 10.15 – Animating the opacity of our dots

Doesn't look that hard, right? We've done the same thing we've done with `dotWidth` but this time, we've created a new variable called `opacity`. We know that the opacity of an element is between `0` and `1`, so we've changed `outputRange` so that it fits our needs.

After that, we introduced our opacity value inside the `style` prop of our `<Animated.View />` component.

Now that we're done with the `Paginator` component, we should implement it inside our `Onboarding.js` file. We know how to do that already: import the component and then place it underneath the `<Block />` component, which has a `flex` of 3 applied to it. Don't forget to pass it the necessary props.

We've learned a lot of things about how animation should work by building this `Paginator` component and for that, I must congratulate you! We've made some impressive progress in this chapter. Now, it's time to start adding some functionality to our screen. Let's learn how we can do that.

Automatic scrolling

To finish building this project, we're going to have to create a button that moves the slides whenever we press it. We already have the `<Button />` component imported from `'galio-framework'`, so let's implement it underneath our `<Paginator />` component:

```
export default function Onboarding() {

  const scrollX = useRef(new Animated.Value(0)).current;
  const viewConfig = useRef({ viewAreaCoveragePercentThreshold: 50 }).current;

  return (
    <SafeAreaView style={styles.container}>
      <Block flex={3}>
        <FlatList
          data={slides}
          renderItem={({item}) => <OnboardingItem item={item} />}
          horizontal
          showsHorizontalScrollIndicator={false}
          pagingEnabled
          bounces={false}
          keyExtractor={(item) => item.id}
          viewabilityConfig={viewConfig}
          onScroll={Animated.event(
            [{ nativeEvent: { contentOffset:  { x: scrollX }}}],
            {
              useNativeDriver: false
            }
          )}
        />
      </Block>
      <Paginator data={slides} scrollX={scrollX} />
      <Button onPress={() => {}} color="#c9a0dc" shadowless>
        Next
      </Button>
    </SafeAreaView>
  );
}
```

Figure 10.16 – The Button component added to our onboarding screen

As you can see, I've implemented `Button` below `<Paginator />`. I've added the same color our images and dots are and removed the shadow via the `shadowless` prop. Now that we know that our function needs to be called whenever we press the button, we need to create a function and then link it to our `onPress` prop.

But before we do that, we need to make sure we have anything in place for our button to work whenever we need it to.

First, we need to think about how we can get to the next slide without swiping through our list of slides. Well, we'll need a reference to the `FlatList` component. Having a reference to that object allows us to control it from an external function whenever we want.

Secondly, we need to keep track of our slides as we need to know which slide we are on at all times. We can do that with a state variable that keeps track of the index that's currently being displayed on the screen.

Now, let's start solving these issues first before we look at what else we need to do to make sure this works.

Let's create a state variable using the `useState` Hook, which we've already imported:

```
const [currentIndex, setCurrentIndex] = useState(0);
```

Here is where we're going to store the index of the currently shown slide.

Now, let's create a ref variable:

```
const slidesRef = useRef(null);
```

Once we've finished creating our ref variable, we should apply it to our `<FlatList />` component. We can do that using `ref={slidesRef}`.

Next, we're going to use a prop that's already available to us from `FlatList` called `onViewableItemsChange`. Whenever you scroll through `FlatList`, the items on `FlatList` will change as well. When those items change, this function is called, telling you what current `viewableItems` are available. This prop should always be used with `viewabilityConfig`. The `onViewableItemsChange` function will be called whenever the corresponding conditions of `viewabilityConfig` are met.

This will help us make sure we always have the right index for our displayed slide. So, inside the function, we'll have to make sure that we set the current index to the one being displayed:

```
const viewableItemsChanged = useRef(({ viewableItems }) => {
    setCurrentIndex(viewableItems[0].index);
}).current;
```

Looking at this might seem a little bit weird but as we discussed earlier, the function will return what current `viewableItems` there are.

The thing is… only one item can be displayed at a time, so the array of `viewableItems` is going to have a single element inside of it. Because what we're interested in is the index of that element, we're setting the `currentIndex` state variable so that it's equal to `viewableItems[0].index`.

Now that we know which slide is currently being displayed, the next step is to just scroll to `currentIndex + 1`. For example, if we're viewing the first slide, that means our `currentIndex` should be equal to `0`. Naturally, the next slide is going to be `currentIndex + 1`, which means `1`:

```
export default function Onboarding() {
  const [currentIndex, setCurrentIndex] = useState(0);
  const slidesRef = useRef(null);

  const scrollX = useRef(new Animated.Value(0)).current;
  const viewConfig = useRef({ viewAreaCoveragePercentThreshold: 50 }).current;

  const viewableItemsChanged = useRef(({ viewableItems }) => {
    setCurrentIndex(viewableItems[0].index);
  }).current;

  const scrollTo = () => {
    if (currentIndex < slides.length - 1) {
      slidesRef.current.scrollToIndex({ index: currentIndex + 1 });
    } else {
      console.log('Last item');
    }
  };

  return (
    <SafeAreaView style={styles.container}>
      <Block flex={3}>
        <FlatList
          data={slides}
          renderItem={({item}) => <OnboardingItem item={item} />}
          horizontal
          showsHorizontalScrollIndicator={false}
          pagingEnabled
          bounces={false}
          keyExtractor={(item) => item.id}
          viewabilityConfig={viewConfig}
          onScroll={Animated.event(
            [{ nativeEvent: { contentOffset:  { x: scrollX }}}],
            {
              useNativeDriver: false
            }
          )}
          onViewableItemsChanged={viewableItemsChanged}
          ref={slidesRef}
        />
      </Block>
      <Paginator data={slides} scrollX={scrollX} />
      <Button onPress={scrollTo} color="#c9a0dc" shadowless>
        Next
      </Button>
    </SafeAreaView>
  );
}
```

Figure 10.17 – The final Onboarding component

Now that we've finished with `viewableItemsChanged` and we've used the variable with our `onViewableItemsChange` prop, let's explain how the `scrollTo` function works.

As you can see, we've created a function called `scrollTo` that is called whenever we press our button. This function checks for `currentIndex` because we want different types of behaviors based on whether we're displaying the last slide or not. If this is the last slide, we won't do anything yet, but if it's the first three slides, we want it to scroll to the next one.

As you can see, scrolling to the next slide is pretty easy – all we have to do is use the reference we have to the `<FlatList />` component and use the `scrollToIndex` function. That function needs an argument where we're telling it which index exactly to jump to.

Now, we can hit **save**, reload our application, and there we have it – a beautiful onboarding screen with some cool little animations and a nice feature that scrolls the slides without us touching anything besides a button. It has been a long journey but I'm pretty sure you think it was worth it, now that we've seen what we're capable of.

In the next chapter, we're going to build the rest of our app but for a nice experience, we're going to use this onboarding screen for our app. This will ensure that at the end of the slides, our button will jump us straight to the app.

Summary

This chapter was one of the hardest challenges we've overcome so far. We've been through so many new concepts but in the end, we can happily say that we've successfully managed to create a great onboarding experience for our users. Even better, we've created a nice onboarding experience that we'll enjoy whenever we brag about our app.

We started by discovering how this app will look and then went through all the necessary steps to produce that app. We saw what it takes to create a nice list of elements, which brought us to `FlatList`. We used this component at the core of our onboarding screen, and you'll surely keep using it in the future whenever you encounter big lists of elements.

We also learned how to create animations and how exactly interpolation works. By doing this, we managed to create a cool little paginator that displays the current slide our users are seeing.

Finally, we even discovered that we can make things work without swiping left or right, just by pressing a button. For this, we used a reference object that is called from another function whenever we pressed that button.

This chapter might have been a lot, but I feel like you were ready for it. I hope you're prepared for the next chapter as well because we're going to finish this mobile app!

11
Let's Build – Stopwatch App

In the previous chapter, we built the beginning of our cool Stopwatch app by creating an onboarding screen. Now, it's time to finish our app by building the other features our users are going to use.

We'll learn lots of new things so that by the end of this chapter, we'll have a pretty cool app that I hope is going to inspire you to create more helpful applications for the rest of the world. The stuff we've learned so far and will continue to learn should give you all the necessary tools to create simple small applications, without the need for another tutorial. Even so, you'll sometimes find yourself looking all over the internet for solutions to your problems, and that's OK. We all do that, so be happy whenever you find a solution and you can make it work.

To build this React Native mobile app, we're going to start by linking our onboarding screen to our actual app by using the React Navigation library. This will help us build the navigation of our screens with little to no effort.

After that, we'll start working on the Stopwatch part of our app. Creating the Stopwatch functionality is pretty straightforward but not as intuitive as you might think.

Once we've created our Stopwatch screen, we will start working on the other part of our app, which is the Timer screen. This will teach us how to play sounds and how to use what we've already learned by creating the Stopwatch app but with a small spin.

Finally, we'll learn about local storage and how to use it to make sure our onboarding screen is not going to show up every time we open the app since it kind of defeats the purpose of having an onboarding screen. So, let's get ready and have some fun coding!

This chapter will cover the following topics:

- Linking to React Navigation
- Creating a Stopwatch
- Creating a Timer
- Finalizing our app

Technical requirements

You can check out this chapter's code by going to GitHub at `https://github.com/PacktPublishing/Lightning-Fast-Mobile-App-Development-with-Galio`. You'll find a folder called `Chapter 11` that contains all the code we've written in this chapter. To use that project, please follow the instructions in the `README.md` file.

Linking to React Navigation

We'll begin this challenge by using the same project we used previously in *Chapter 10, Building an Onboarding Screen*. Why, you ask? Well, that's because the purpose of creating an onboarding screen was exactly this – to have some sort of introduction to our main app.

So, open the folder and get ready to code. We'll start by importing all the necessary packages we'll need to connect our onboarding screen to any new screen we'll be creating moving forward. Let's open our terminal and move to our project folder. There, we'll begin by writing the following command:

```
npm install @react-navigation/native
```

This will install the basis of our navigation system. We'll also need to install all the dependencies this package needs, which we can do via the following command:

```
expo install react-native-gesture-handler react-native-
reanimated react-native-screens react-native-safe-area-context
@react-native-community/masked-view
```

Now that all our dependencies have been installed, let's talk about the **React Navigation** library.

There are several options for people trying to create a navigation system with React Native, but the most commonly used one is React Navigation. You might be wondering why that is, and my answer for you would be that this is the most maintained and packed features library of them all. I strongly recommend diving into their documentation, which you can find at https://reactnavigation.org/.

On top of being such a good navigation library for React Native, it also has a really easy and straightforward way of setting up your routes, which we'll look at later in this chapter. So, on top of being easy to use, it's completely customizable and has native support for iOS and Android. What more could you ask for from a navigation library?

Let's move on with our app and think about what everything should look like. I was thinking that once the user is finished with our onboarding screen and they hit the **Next** button for the last time, our user will be transported to another screen, straight to the **Stopwatch** screen. This screen will have two tabs: one for the Stopwatch, which is the main use case of our app, and another for the **Timer** screen.

For that to work, we'd need two new components from @react-navigation: stack and bottom-tabs. Let's import them with the following commands:

```
npm install @react-navigation/stack
```

Now, it's time for us to install the next package we'll be using:

```
npm install @react-navigation/bottom-tabs
```

Now that everything has been installed, it's time for us to restructure our project so that we have better control over where our files go.

We'll create a new folder inside the root directory of our project called screens. Next, we must copy and paste our Onboarding.js file from the components folder.

Once you've moved that file into the proper directory, it's time to check our files to make sure they all link to this new path we have for our onboarding screen. We also need to see if there are any imports inside Onboarding.js that need to be modified.

Our imports from inside `Onboarding.js` that need to be changed are for the components we're using inside this screen: `OnboardingItem` and `Paginator`. Because those components are not in the same folder anymore, we'll have to make sure they're imported with the correct path. In our case, the path changes to `"../components/OnboardingItem"`:

```
import OnboardingItem from '../components/OnboardingItem';
import Paginator from '../components/Paginator';
```

Figure 11.1 – Our new imports for the onboarding screen

Because we're already here, just go to the `scrollTo()` function. Instead of the `console.log()` line that we have inside the `else` statement, write the following line:

```
navigation.navigate('Tab Navigator');
```

This is telling `Button` that once it gets to the end of the onboarding screen, the next step is to navigate to the next screen, called `'Tab Navigator'`. We'll introduce this screen when we create our routing system. Because we're using a variable called `navigation`, we should also let our component know where to get it. Directly above where we're defining our `Onboarding` function and between the parentheses, we'll allow our function to receive this prop, called `navigation`, like this:

```
export default function Onboarding({ navigation }) {
```

Now, if we want to have a working app, we'll have to go to `App.js` and change the import for the onboarding screen to the correct path as well. Once we've finalized the changes with the correct imports, we can save and run the app. Nothing should have changed; all we did was add a new directory so that we have a better folder structure. Some text editors or IDEs will automatically change the imports for you, so make sure you always read whatever messages might pop up.

> **Tip of the day**
>
> I often refresh my app and check for changes or error messages, especially when all the changes inside the app shouldn't change anything visually. This way, I can make sure I'm always up to date with whatever happens inside the app when it rerenders.

Now that we have a new folder structure, we can start working on creating the routes needed for our app to work. But first, we need to have some placeholders for the screen we're going to work with. So, let's create two new files in our `screens` folder: `Stopwatch.js` and `Timer.js`.

For both, we'll have the same code, besides the name of our functions, which will be written inside the <Text /> component. We'll need those files to test if our routes are working correctly before we start diving into the functionality of our app.

Let's see what that placeholder screen looks like:

```
import React from "react";
import { View, Text, StyleSheet } from "react-native";

export default function Stopwatch() {
  return (
    <View style={styles.container}>
      <Text>Stopwatch</Text>
    </View>
  );
}

const styles = StyleSheet.create({
  container: {
    flex: 1,
    backgroundColor: "white",
    alignItems: "center",
    justifyContent: "center",
  },
});
```

Figure 11.2 – Placeholder screen for testing out routes

This example was specifically created for the Stopwatch.js file. You'll have to create a second one for Timer.js as well. As I've already specified, the difference between this one and the Timer one is going to be in the name of the function and whatever's written inside the <Text /> component. The rest of it should be the same as we're only using these files to test out our routes.

Now that we have these two new files inside our screens folder, we can go ahead and create a new file in our root directory called routes.js. This is where we're going to create the routing system for our cool little app.

Once you've created the new file, we can open it and start coding. We'll begin by importing all the necessary packages and files that we're going to need for this routing system. You can see what packages I'm importing by looking at the following screenshot:

```
import React from 'react';
import { NavigationContainer } from '@react-navigation/native';
import { createStackNavigator } from '@react-navigation/stack';
import { createBottomTabNavigator } from '@react-navigation/bottom-tabs';

// screens
import Onboarding from './screens/Onboarding';
import Stopwatch from './screens/Stopwatch';
import Timer from './screens/Timer';
```

Figure 11.3 – Imports for routes.js

Now, as you can see, we've been importing all the main packages from `@react-navigation`. We started by importing React as we need it to create this component-based routing system. Next in line is the `NavigationContainer` component, which was imported from `@react-navigation/native`. This component deals with managing the app's navigation state and creating the connection between your top-level navigator and the app environment.

After this, we imported `createStackNavigator` and `createBottomTabNavigator`. To understand how the **Stack Navigator** works, we'd have to start thinking of our screens as cards in a deck of cards. You're always placing a new card on top of an old card so that you can create a stack of cards. That's basically how React Navigation works, always placing a new screen on top of another screen.

The **Bottom Tab Navigator** creates the common bottom bar you usually come across whenever an app wants you to have easier access to the main functionality. This way, we can let our user quickly switch between the Timer and Stopwatch, with each screen having easy access from the bottom bar.

Once we've imported the necessary dependencies to create a routing system for our app, it's time to import the screen we'll be using in this system. Of course, the onboarding screen is really important as this must be the first screen our users see, after which we need the Stopwatch and Timer screens.

Now that we're done with the imports, it's time to see how we can use React Navigation to create our routing system. We'll use `createStackNavigator` and `createBottomTabNavigator` to create the variables that we're going to use as components for defining our screens and navigators, so let's do that now:

```
const Tab = createBottomTabNavigator();
const Stack = createStackNavigator();
```

Figure 11.4 – Creating variables out of our navigation functions

Having these variables enables us to create easy-to-read routing systems.

Let's begin by writing the function for our main screens; that is, Stopwatch and Timer. This should be a normal React function that returns the system for a Bottom Tab Navigator. So, we'd use the Tab variable for this. Let's see what our function looks like:

```
function AppTabs() {
    return (
        <Tab.Navigator>
            <Tab.Screen name="Stopwatch" component={Stopwatch}/>
            <Tab.Screen name="Timer" component={Timer} />
        </Tab.Navigator>
    );
}
```

Figure 11.5 – Main screen routing for the Stopwatch and Timer screens

This looks pretty easy to understand, right? We have a <Tab.Navigator /> component that has two screens using the <Tab.Screen /> component as children. The Navigator component acts like the glue that lets React Native know that those two screens need to be part of the Bottom Tab Navigator.

For every routing system like this, we need a Navigator component and then some Screen components that let Navigator know which screens are part of it.

I feel like this is pretty straightforward to use in that anybody could just go ahead and start creating routing systems for their apps. I encourage you to use routing inside your apps as much as possible, just to see how many options and things you can change. React Navigation is extremely customizable, so I'm pretty sure you'll be amazed by the possibilities of using this library.

Now, the next step is to set up our main stack of screens. We'll do that the same way we've set up our `AppTabs()` function component but this time, we'll also use the `<NavigationContainer />` component as this will be our main routing component:

```
export default function AppStack() {
    return (
        <NavigationContainer>
            <Stack.Navigator>
                <Stack.Screen name="Onboarding" component={Onboarding} options={{
                    headerShown: false
                }} />
                <Stack.Screen name="Tab Navigator" component={AppTabs} options={{
                    headerShown: false
                }} />
            </Stack.Navigator>
        </NavigationContainer>
    );
}
```

Figure 11.6 – Main routing system for our app

Looking at this code of our main function for our routing system might make you ask yourself what's going on here. Fear not – this is not that hard to understand. Because this is going to be our main routing system, we've used the `<NavigationContainer />` component. Inside of it, we have a `<Stack.Navigator />` component creating a set of screens that can be applied one on top of each other, just like a deck of cards. Here, we have two screens: the `Onboarding` screen and the `AppTabs` screen.

As we saw earlier, we've defined the `AppTabs` screen component as a Bottom Tab Navigator screen containing our two main screens: `Stopwatch` and `Timer`.

We also have a `prop` called `options` applied on both of our `<Stack.Screen />` components. This prop allows us to apply custom characteristics to our screens. Because React Native enables a header bar on each screen in a stack by default, we had to get rid of it, so we've given it a value of `false`. If we didn't specify this, every time you go to this screen, you'd see the default platform header at the top of the screen.

Now that we have exported this function, we can go inside our `App.js` file and apply our routing system. But this file is filled with tons of stuff we don't need, so let's clean it up. Delete everything in `App.js` so that we can start rewriting it in the best way possible for our use case.

After removing everything from inside the file, we can start by importing React. After that, import the `AppStack` component we defined earlier in the `routes.js` file. Now, all we have to do is create a function called `App()` that returns our `<AppStack />` component, as shown in the following screenshot:

```
import React from 'react';

import AppStack from './routes';

export default function App() {
  return (
    <AppStack />
  );
}
```

Figure 11.7 – The App.js file after making all the necessary modifications

Now, our `App.js` file looks a lot cleaner, and we've successfully connected our routing system to our React Native app. You should test your app! Save everything, start the Expo server, and open your preferred simulator or physical device.

Because we've already linked the onboarding screen to the Tab Navigator screen in the `scrollTo()` function via the `navigation.navigate()` function, we now have a fully functional routing system.

You should now be able to see the onboarding screen first. Hit the **Next** button until you get to the last screen. Once you are there, hit **Next** one more time, and boom! You're now in the `AppTabs()` Tab Navigator. That's the component we've been defining in our `routes.js` file. You can click the **Bottom Tab Navigator** button to quickly switch between the Stopwatch and Timer apps.

Our React Navigation implementation was a success! Now, it's time for us to start coding the functionality of our Stopwatch screen.

Creating a Stopwatch

Some of you that have already worked a bit with JavaScript might think that creating a Stopwatch is as easy as just calling the `setInterval()` function and subtracting a number at every iteration. Well, not really, but fear not – we'll make this as easy as possible for everyone, regardless of your experience with JavaScript.

So, let's begin by opening our `Stopwatch.js` file, which we can find inside the `screens` folder. Right now, there's only some text with the word Stopwatch inside of it that is centered because we styled the main `<View />` component.

I'd honestly just start by removing everything from this file and starting fresh with the imports. We'll begin by importing `React`, `useState`, and `useEffect` from `'react'`. After that, we'll import `StyleSheet` and `SafeAreaView` from `'react-native'`. Finally, we will import the `Text`, `Block`, and `Button` components from `'galio-framework'`.

After importing the components that we're going to use to create this screen, it's time to build a static screen for us to serve as a starting point. Let's take a look at the following code and try to explain it as this is going to be our main layout skeleton:

```jsx
import React, {useState, useEffect} from 'react';
import { StyleSheet, SafeAreaView } from 'react-native';
import { Text, Block, Button } from 'galio-framework';

export default function Stopwatch() {
    return (
        <Block style={styles.container}>
            <SafeAreaView style={{ flex: 1 }}>
                <Block flex={0.32} alignItems="center">
                    <Text style={{ fontSize: 72, marginTop: 32 }}>00:00.00</Text>
                    <Block row space="around" style={{ width: '100%' }}>
                        <Button size="small" shadowless>Start</Button>
                        <Button size="small" shadowless>Lap</Button>
                    </Block>
                    <Block row justifyContent="center" alignItems="center" style={{ width: '100%' }}>
                        <Block style={styles.divideLine} />
                        <Text size={16}>Laps</Text>
                        <Block style={styles.divideLine} />
                    </Block>
                </Block>
                <Block flex={0.68}>
                    <Text>Laps showing here</Text>
                </Block>
            </SafeAreaView>
        </Block>
    )
}

const styles = StyleSheet.create({
    container: {
        flex: 1,
        backgroundColor: 'white',
    },
    divideLine: {
        borderWidth: 1,
        borderColor: '#6c757d',
        margin: 10,
        flex: 1
    }
});
```

Figure 11.8 – Basic layout of our Stopwatch component

Well, this is a big chunk of code, so let's dive right in and explain it. So, after importing everything that we need, we will begin writing our Stopwatch() functional component. Inside of it, we can see that there's a big <Block/> component and then a <SafeAreaView /> component. These are here just to take everything in and make sure there won't be any problems if we encounter a phone with a notch.

Up until now, everything has been really easy, so what happens next? We must split the screen into two <Block /> elements, one with a flex property of 0.32 and the other with a flex property of 0.68. This is so we can make sure we'll have an upper side of the screen containing all the buttons and functionality, and then a lower part of the screen where all the laps will appear.

On the upper side of the screen, we can see that we have a `<Text />` element with a big font size. This will be our time, which is going to be changed when we add all the functionality. After that, we have another `<Block />` element with the row prop enabled. This has two buttons inside it. We'll use those buttons to start/stop the Stopwatch and also create laps whenever someone has finished a lap.

After that, we have another `<Block />` element, whose purpose is to make our layout a bit more intuitive to the users. It will point out that the laps will show up underneath that line. We've created some styling for those lines that you can find in the styles object under `divideLine`.

The following screenshot shows what this will look like on our devices:

Figure 11.9 – Basic layout of our Stopwatch screen

Nice! Now that we have the basic layout coded out, it's time to start working on the functionality of our screen. We should start by defining some state variables that we're going to use all over this screen. But before that, let's go back to the beginning for a second and think about why I said that we can't increment the time with the `setInterval()` function.

Using setInterval

So, `setInterval` is a function that does exactly what you'd expect it to do. You set a time interval such as 1,000 milliseconds, which is one second, and for every second, a function that you're going to define is going to be called. You might think that in theory, we can have something like the following for our Stopwatch screen's functionality:

```
let ms = 0;

setInterval(() => {
    ms += 10;
}, 10);
```

Figure 11.10 – setInterval used in a basic example

This would work pretty well. Here, every 10ms, we're firing the function that increments our variable with 10. In theory, this works great as we'd now have a basic Stopwatch built in five lines of code. But the thing is, `setInterval()` is not that reliable.

Why am I saying this? Well, if you look at the preceding function, we can see that we've specified 10ms as the timing parameter, so our function should fire up every 10ms. However, it will not execute the code at the specified time. Instead, it will wait *at least* 10ms before it executes. Because of that, we can't say our time function will be accurate.

I tried a different workaround and I figured out that the best way we can handle the time is by using `Date` objects.

Now that we've figured this out, let's write our state variables:

```
// controls
const [startTime, setStartTime] = useState(); // the first value setting the basis of our time
const [laps, setLap] = useState([]); // array of laps
const [started, setStarted] = useState(false); // if the clock has started
const [elapsed, setElapsed] = useState(); // time passed in ms
const [intervalId, setIntervalId] = useState(); // allows us to clear the interval whenever we want

// time
const [minute, setMinute] = useState(0);
const [seconds, setSeconds] = useState(0);
const [ms, setMs] = useState(0);
```

Figure 11.11 – State variables used inside the Stopwatch component

I've explained each of them but basically, we'll have five controls state variables called `startTime`, `laps`, `started`, `elapsed`, and `intervalId`. Then, we have the time state variables, which we're going to use to see the time change on the screen. These are called `minute`, `seconds`, and `ms`.

Now, let's use the time state variables and make them visible on the screen. Let's see what that `<Text />` component looks like now, after applying the time state variables:

```
<Text style={{ fontSize: 72, marginTop: 32 }}>{minute < 10 ?
`0${minute}` : minute}:{seconds < 10 ? `0${seconds}`: seconds}.
{ms}</Text>
```

Because we might have single-digit numbers at one point in time, by writing our variables like this, we can make sure that if they're single-digit numbers, we'll add one `0` at the beginning. We will do this for both the `minutes` and `seconds` variables.

Saving the file and refreshing the screen should show you no changes at all. This is good as it means we've implemented the time state variables inside our `Stopwatch` component correctly.

Now that those variables are in place, let's build a function that will be called once we press the **Start** button on the screen. This button needs to do several things; first of all, it needs to function as a **Start** and a **Stop** button. Secondly, it needs to initialize the `startTime` control state variable we've just defined with a new `Date` object. Let's take a look at this function now:

```
const startAndStop = () => {
    if(!started) {
        setStartTime(new Date().valueOf());
        setStarted(true);
    } else if (started) {
        setStarted(false);
    }
};
```

Figure 11.12 – The startAndStop() function

So, this function can do two things. First, if our `started` state variable is `false`, we'll set it to `true` to announce the start of the Stopwatch and then set the `startTime` variable to a `new Date()` object. By having a date set at the beginning, we can use it later to calculate how much time has passed between each iteration, allowing us to have a more accurate time displayed.

Now, once the started variable has been changed, we need to start the setInterval() function. Let's take a look at the following function and talk about how it works:

```
useEffect(() => {
    let _intervalId;
    if(started) {
        _intervalId = setInterval(() => {
            setElapsed(new Date().valueOf());
        }, 10);

        setIntervalId(_intervalId);
    }
    if(!started && intervalId !== undefined) {
        clearInterval(intervalId);
    }

    return () => {
        if (intervalId) {
            clearInterval(intervalId);
        }
    };
}, [started]);
```

Figure 11.13 – The useEffect() implementation for starting the setInterval() function

We've implemented this useEffect() function because React gave us this cool little function that is called every time the component rerenders. The coolest thing about it is that we can condition it to work only when the state variable in the second parameter is changed. Every time the started variable is changed, this function is getting called.

The function, which is called once the started variable is changed, will be the core of our Stopwatch functionality. This functionality will be inside the setInterval() function. Inside that function, we're setting our elapsed variable to a new Date() *every 10ms*. After that, we're grabbing our setInterval() function and applying it to the intervalId state variable.

The return function inside the useEffect() function cleans up after a side effect. This means that this function will be called every time the started variable changes, just to clean up after the previous render. It also gets called when the component unmounts. Because we're working with setInterval(), I wanted to make extra sure that our interval is going to be cleared every time our started variable is set to false (the Stopwatch stops) so that it won't weigh heavily on a user's CPU.

As you can see, clearing an interval is as easy as calling the clearInterval() function and passing it the interval we want to clear.

Now that we have the latest time in our elapsed variable, all we have to do is check out the difference between the `elapsed` time and `startTime`. We can do this with `useEffect()`. Every time the `elapsed` state variable is changed, another `useEffect()` function will trigger. Then, we can do all the math there. Let's take a look at how I've done this:

```
useEffect(() => {
    if(elapsed && started) {
        let delta = Math.abs(elapsed - startTime) / 1000;

        let minutes = Math.floor(delta/60) % 60;
        delta -= minutes * 60;

        let seconds = Math.floor(delta % 60);

        let totalMs = Math.floor((elapsed - startTime) / 10) % 100;

        setMs(totalMs);
        setSeconds(seconds);
        setMinute(minutes);
    }
}, [elapsed]);
```

Figure 11.14 – The second useEffect() function, which depends on elapsed

So, this `useEffect()` function is called every time `elapsed` changes, which is every 10ms. What we're doing here is checking if `elapsed` exists (is not undefined) and that `started` is `true`. If both these conditions are true, we can use `elapsed`, which contains the latest `Date` value, to work out the difference and have it in milliseconds. Moving forward, we do the math for the minutes and seconds. Once we have all these values, we can set them to the time state variables we defined earlier: `minute`, `seconds`, and `ms`.

Wait… are we done? Kind of, but not really. Let's go to our **Start** button and change it a little bit so that it can be used. We'll change it like this:

```
<Button size="small" color={started ? "#6c757d" : "#c9a0dc" }
onPress={() => startAndStop()} shadowless>{started ? "Stop" :
"Start"}</Button>
```

This way, we can have different colors and our button will display a different text based on whatever it can do at that moment. This is all based on our `started` state variable, which tells us if the Stopwatch has started or not. I've picked these colors because we've used them in the onboarding screen as well and I thought they fit, but you can use whatever colors you want.

Now, we can save and reload our JavaScript and check out what we've been creating. By pressing the **Start** button, you'll see it change its text into the **Stop** button's text, and that its color is now gray. The time started going up and our Stopwatch works correctly, but what type of Stopwatch is this if we can't even register any laps?

Let's create the function that'll be busy registering the laps. We also need a way to display those laps, which we'll do once we're done with the function. I was thinking that we can use this function just like we used the startAndStop() function, in that we should be able to register laps and clear all the laps with the same button. So, let's take a look at how I did this:

```
const lap = () => {
    if(started)
        setLap(oldArray => [...oldArray, [minute, seconds, ms]]);
    else
        setLap([]);
}
```

Figure 11.15 – setLap function used to register and clear all the laps

This is a straightforward function; our function can do two different things based on what our started state variable value is. If the Stopwatch is started, it's OK for us to register new laps but if the Stopwatch is not working anymore, we should be able to clear all the laps and get ready for a new session.

Now that we have this function, let's link it to our **Lap** button, just like we did with the **Start** button:

```
<Button size="small" color="#f4d1dc" onPress={() => lap()}
shadowless>{started ? "Lap" : "Clear laps"}</Button>
```

Now, let's work on displaying the laps on the screen. We're going to do that by importing FlatList from 'react-native', so just scroll up to the imports part of our file and add FlatList. Our new import should look like this:

```
import { StyleSheet, SafeAreaView, FlatList } from 'react-native';
```

Displaying laps

We should use the `FlatList` component in the `<Block flex={0.68} />` component instead of the `<Text />` component currently living there. So, delete the `<Text />` component and look at my implementation of `<FlatList />`:

```
<Block flex={0.68}>
    <FlatList
        data={laps}
        renderItem={(({item}) => <LapItem item={item} />}
        bounces={false}
        showsVerticalScrollIndicator={true}
        keyExtractor={(item) => laps.indexOf(item).toString()}
    />
</Block>
```

Figure 11.16 – FlatList implemented instead of our Text component

There's nothing new here. We've already used a `FlatList` component to build the onboarding screen, and you've probably noticed that we have a new component there called `<LapItem />`. I've defined this component under our main Stopwatch component. You could always move it and create a separate file for it under the `components` folder, but I felt like having it close to the main component was easier for me to always check out. Let's check out this component:

```
const LapItem = ({ item }) => {
    return (
        <Block row fluid justifyContent="center">
            <Text h3>
                {item[0] < 10 ? `0${item[0]}` : item[0]} : {item[1] < 10 ? `0${item[1]}` : item[1]}.
            </Text>
            <Text p style={{ alignSelf: 'flex-end' }}>
                {item[2] < 10 ? `0${item[2]}` : item[2]}
            </Text>
        </Block>
    );
};
```

Figure 11.17 – The LapItem component being used in FlatList

This component takes a `prop` called `item`, which is an array containing the information needed to display the laps.

And with that, we've finished this beautiful screen. Save and reload the JavaScript and try it out. The following screenshot shows what our app should look like right now:

Figure 11.18 – Completed Stopwatch component

It's working! It looks cool and we've had a great time building this. Now, let's start working on the Timer screen.

Creating a Timer

Now that we're done with the Stopwatch screen, it's time to open the `Timer.js` file and work on our Timer screen. We should dive right in, remove everything, and start importing everything we need for this.

First of all, we all know that at the end of a Timer cycle, there's always a sound playing, letting you know that it has stopped. For that, we need a new package called `expo-av`. This is an Expo package for dealing with audio. It is quite easy to use, so let's begin by installing it into our project by using the following command:

```
expo install expo-av
```

Now that we have installed this, we can start importing everything we'll need to build this component. We'll need a timer component that is quite similar to the Stopwatch. We'll also need to use intervals and date objects to calculate everything, so `useEffect` and `useState` will be imperative for our screen.

The difference is that we'll need to let the user input exactly how much time they want the Timer to work for. We can do this with a `<TextInput />` component from `'react-native'`. Because we're using an input, we'll also need a `<KeyboardAvoidingView />` component that helps us reorganize the layout so that our inputs will never be hidden by the opening of the keyboard. Let's take a look at our imports:

```
import React, {useState, useEffect} from 'react';
import { StyleSheet, TextInput, KeyboardAvoidingView, SafeAreaView } from 'react-native';
import { Audio } from 'expo-av';
import { Text, Block, Button } from 'galio-framework';
```

Figure 11.19 – Imports for the Timer screen

As you can see, the `import` statements are quite similar to the Stopwatch screen. This is because these screens are similar. However, by having them doing the same thing, we can learn to always inspire ourselves by looking at what we did in the past. All the code you've written will help you with other problems you might encounter. Because we've created the Stopwatch screen, we now know of the inaccuracy of `setInterval()` and how to combat that.

Now, let's start creating the basic functionality for our Timer screen with a layout that we can work with. For the layout, we'll start everything the same way we've started the Stopwatch screen; that is, with a `<Block />` component with `styles.container` attached to it. After that, we'll use `<SafeAreaView />` and then a `<KeyboardAvoidingView />` with a `flex: 1` style applied to it.

Inside of that `<KeyboardAvoidingView />` component, we'll have two `<Block />` elements. The first one is going to have a `<TextInput />` component as a child that is going to be the title of the Timer. We're using `<TextInput />` here because you may wish to change the title of the timer; this is just a cool little feature. The second one is going to have two `<TextInput />` elements – one for minutes and one for seconds. This way, the user can input whatever number they need for the timer. The second `<Block />` element will also contain the **Start/Stop** button of the timer. Let's see what that looks like:

```
return (
    <Block style={styles.container}>
        <SafeAreaView style={{ flex: 1 }}>
            <KeyboardAvoidingView style={{ flex: 1 }}>
                <Block flex style={{ paddingHorizontal: 16 }}>
                    <TextInput
                        value={title}
                        onChangeText={setTitle}
                        returnKeyType='done'
                        clearTextOnFocus
                        style={{ borderBottomWidth: 1, fontSize: 54 }}
                    />
                </Block>
                <Block flex alignItems="center">
                    <Block row>
                        <TextInput
                            style={{ borderBottomWidth: 2, fontSize: 54 }}
                            keyboardType="numeric"
                            value={countdownMinutes}
                            onChangeText={setCountdownMinutes}
                            returnKeyType='done'
                            editable={!startTimer}
                            clearTextOnFocus
                        />
                        <Text style={{ alignSelf: 'center', fontSize: 54 }}>:</Text>
                        <TextInput
                            style={{ borderBottomWidth: 2, fontSize: 54 }}
                            keyboardType="numeric"
                            value={countdownSeconds}
                            onChangeText={setCountdownSeconds}
                            returnKeyType='done'
                            editable={!startTimer}
                            clearTextOnFocus
                        />
                    </Block>
                    <Button size="large" onPress={() => start()}
                    style={{ marginTop: 32 }} color={startTimer ? "#6c757d" : "#c9a0dc"}
                    shadowless>
                        {startTimer ? "Stop" : "Start"}
                    </Button>
                </Block>
            </KeyboardAvoidingView>
        </SafeAreaView>
    </Block>
)
```

Figure 11.20 – Layout created for the Timer screen

As we explained earlier, this is not that complicated, but what you'll notice here is that I've already filled in the value props for our `<TextInput />` components. I've also made sure that a state variable is in place for our **Start/Stop** button. This is only because we've already been through the Stopwatch screen, which means we've already experienced the fact that we'll need certain state variables so that we can change the text inside the buttons.

As you can see, we also used the `editable` prop on our `<TextInput />` components since we only want those values to be **editable** when the timer is not working. We can also see another new prop, called `returnKeyType`. This prop allows us to tell the device which kind of key we want available for our users. I've opted for the `done` key because once they add the number they want, they could just press that key and move forward.

We also know from the previous chapters that `<TextInput />` is a controlled component, which means it needs a state variable for the `value` prop while also a way of changing that state via the `onChangeText` prop. Knowing all this, I'd suggest reading over that code a couple of times to see if you can understand it. We won't make any changes to it anymore as this is enough for us to be able to jump right into the Timer's functionality.

Let's look at the state variables we've defined for this Timer:

```
// time
const [countdownMinutes, setCountdownMinutes] = useState("00");
const [countdownSeconds, setCountdownSeconds] = useState("30");

// controls
const [final, setFinal] = useState(); // time in future that the timer needs to get to
const [timer, setTimer] = useState();
const [timeDisplay, setTimeDisplay] = useState();
const [intervalId, setIntervalId] = useState();
const [startTimer, setStartTimer] = useState(false);

const [title, setTitle] = useState('Exercise 01');

const [sound, setSound] = useState();
```

Figure 11.21 – State variables created for the Timer screen

So, at the beginning of our functional component for the Timer screen, we'd be writing all these state variables. Just like before, we have some time state variables and some controls' state variables. We've already seen `intervalId` and `startTimer` (we called it `started` in the Stopwatch screen).

Let's briefly explain the other state variables that we're using. The `countdownMinutes` and `countdownSeconds` variables are strictly used to display the minutes and seconds of our timer. The `final` variable is there because we'll know how much time we want our Timer to last based on the user's input. We'll use it to calculate the amount of time once we start the timer.

As you can see, we also have the `timer` and `timeDisplay` variables. The `timer` variable acts just like the `elapsed` variable acted when we were working with the Stopwatch screen. The `timeDisplay` variable is here for us to always have the value of our timer in seconds. This way, we can make sure to stop it whenever it hits 0.

Our `title` variable is there for the title of the screen, which the user will be able to change whenever they want. The sound variable is there because we need to know whenever the sound has been loaded on the screen. This will help us use another `useEffect()` function so that we can clean up after ourselves.

Great! I like how much faster we can move now that we're done with the Stopwatch screen. This means that we're learning, and experience is the best teacher out there! Let's take a look at the `start()` function, which we're going to call whenever we press the **Start** button:

```
const start = () => {
    if(!startTimer) {
        var t = new Date();
        t.setMinutes(t.getMinutes() + parseInt(countdownMinutes));
        t.setSeconds(t.getSeconds() + parseInt(countdownSeconds));

        setFinal(t.valueOf());
        setStartTimer(true);

        loadSound();
    } else if(startTimer) {
        setStartTimer(false);
        clearInterval(intervalId);
    }
};
```

Figure 11.22 – The start() function used to start or stop the timer

As we can see, we're following the same pattern we used for the Stopwatch screen. This enables us to use this function as a start or stop function for our button. So, if the `startTimer` state variable is `false`, then we will initialize a new variable with a `Date` object. Then, we'll set that date with the minutes and seconds we've grabbed from our two `<TextInput />` components from the screen, adding those to the current date's minutes and seconds. This means that we've taken the current date and added the time our user has inputted. This is the final date we're trying to reach, so the next step is setting our `final` state variable with the date we've just calculated.

Then, we must set the `startTimer` variable to true, which will notify our component that the timer has started. At this point, we'll also load the sound. Let's define the `loadSound()` function for this:

```
const loadSound = async () => {
    const { sound } = await Audio.Sound.createAsync(require('../assets/sounds/ring.mp3'));

    setSound(sound);
}
```

Figure 11.23 – The loadSound() function with a new keyword called async

As you might have figured out by now, this function has a new keyword called `async`. Worry not – this is here to make sure our function will not stop the whole application while we're trying to load that sound. If you don't have a sound that you can use, you can find the sound that I've created in this project's files on GitHub. You can also create your own sounds or even use something online that isn't copyrighted. I followed the `'expo-av'` documentation to load the sounds. That's why I'm always stressing the fact that your first step, whenever there's something you don't understand, should be to look at the documentation of that specific package/library.

Now that we've loaded the sounds and started our timer, we should be able to see where all the logic behind our screen exists. Just like we did previously, we're using `useEffect()` functions to make sure they're only triggered when a certain state variable changes. The first `useEffect()` function we're going to use is going to depend on the `final` state variable. That's because this variable is needed for all the math, so, naturally, we're going to check for it before we do anything else:

```
useEffect(() => {
    let _intervalId;
    if (final) {
        _intervalId = setInterval(() => {
            setTimer(new Date().valueOf())
        }, 250);
    }
    setIntervalId(_intervalId);
    return () => {
        clearInterval(_intervalId);
    }
}, [final]);
```

Figure 11.24 – The first useEffect() function depending on the final variable

So, this is exactly like what we did previously. Here, we have a `setInterval()` function being called – this time, every 250ms – but only if the final variable has been initialized. By this, we mean if the user has pressed the **Start/Stop** button. There's nothing weird going on down here, so I think we should be able to move on and check out the following `useEffect()` function:

```
useEffect(() => {
    if (final && timer) {
        let display = Math.floor((final - timer) / 1000);
        if (timeDisplay !== display)
            setTimeDisplay(display); // total seconds

        const minutes = Math.floor(display / 60) % 60;
        display -= minutes * 60;
        setCountdownMinutes(minutes < 10 ? `0${minutes}` : minutes.toString());
        setCountdownSeconds(display < 10 ? `0${display}` : display.toString());
    }
}, [timer]);
```

Figure 11.25 – The second useEffect() function depending on the timer variable

Because we're setting the `timer` state variable in the first `useEffect()` function, it will act as a trigger for the second `useEffect()` function so that it's called. This is the function where we're doing all the math necessary for our timer to work. We're calculating the difference in seconds between the final date and the new date we've received from the `timer` variable.

After calculating the difference in seconds, the next step is to check if the `timeDisplay` variable is different than the newly calculated difference. If it's different, we'll set `timeDisplay` to this new value. We're doing this to make sure we're always calculating everything with a new value.

After that, we just do the usual math for calculating the minutes and seconds. Next, we must set our `<TextInput />` components via the `countdownMinutes` and `countdownSeconds` variables to our newly calculated values. The reason we've used `toString()` here is because a `<TextInput />` component only accepts strings as values, so with this function, we're transforming the value from a number into a string.

Now, let's check out the following `useEffect()` function, which depends on the `timeDisplay` state variable:

```
useEffect( () => {
    if (timeDisplay <= 0) {
        clearInterval(intervalId);

        setTimeDisplay(0);

        setCountdownMinutes("0");
        setCountdownSeconds("0");

        setStartTimer(false);

        if (sound)
            sound.playAsync();
    }
}, [timeDisplay]);
```

Figure 11.26 – The third useEffect() function depending on the timeDisplay variable

This one will only work if `timeDisplay` is 0 or less. Once we hit 0, we should stop and reset all the variables we've been using until now. This is where we're making sure our interval will be cleared, our `timeDisplay` variable goes back to 0, and that the countdown variables go back to 0. This should also be where we play our sound, so we must check if that sound has been loaded and then use the `playAsync()` function to start it.

Because we've been loading the sound when our timer application starts, we should also unload it when the screen unmounts. Practically speaking, there is no reason for us to keep a sound in the memory of our device if we're not using it. We will do this by using a cleanup function inside another `useEffect()` function. Let's check out what that looks like:

```
useEffect(() => {
    return sound ? () => {
        sound.unloadAsync();
    } : undefined;
}, [sound])
```

Figure 11.27 – The fourth useEffect() function being used only for the cleanup function

So, yet again, this is directly inspired by their documentation. Once the component unmounts, we must call this function, which checks if the sound has been loaded. If it has, then we must call the `unloadAsync()` function on our `sound` state variable.

Congratulations! We're done with the Timer screen! Let's check out how it looks and if it works:

Figure 11.28 – Completed Timer screen

It looks great! Here, we can change the title; we can also change the timer's values and that it works once we hit start. Upon waiting for 30 seconds, a cool little sound will play!

With that, we're done with this app! Wait… not really – the Bottom Tab Navigator looks kind of empty, so we should add some icons. There are also some small things we can still do to enhance the user experience. Let's move on and start adding all these little enhancements.

Finalizing our app

At this point, we need to add some icons to the Bottom Tab Navigator.. But how should we do this? Luckily for us, React Navigation has a pretty straightforward way of modifying its default components.

And because we're already here, we should also change the focus color of whatever tab we're using at the moment.

So, let's go back to our `routes.js` file. Because we want to add icons to our tabs, we should import the `Icon` component from `'galio-framework'`. After all the imports, we should write the following:

```
import { Icon } from 'galio-framework';
```

Now that we've imported the component that we'll use to display icons, let's see how we should do that. Search your `AppTabs()` function and find the `<Tab.Navigator />` component. Here, we'll add two new props called `screenOptions` and `tabBarOptions`. Let's quickly check them out and see how exactly we're using them to implement an icon inside our Bottom Tab Navigator:

> **Important Note**
>
> As of August 14, 2021, React Navigation has been updated to v6 and `tabBarOptions` was deprecated in favor of the `options` prop, which works on a per-screen basis. For more information regarding versioning and the React Navigation library, I'd suggest reading the documentation, which can be found at `https://reactnavigation.org/`.

```jsx
function AppTabs() {
    return (
        <Tab.Navigator
        screenOptions={(({route}) => ({
            tabBarIcon: ({ focused, color, size }) => {
                let iconName;

                if (route.name === 'Stopwatch') {
                    iconName = focused ? 'stopwatch' : 'stopwatch-outline';
                } else if ( route.name === "Timer") {
                    iconName = focused ? 'timer' : 'timer-outline';
                }

                return <Icon name={iconName} family="ionicon" size={size} color={color} />
            },
        }))}
        tabBarOptions={{
            activeTintColor: '#c9a0dc',
            inactiveTintColor: '#6c757d',
        }}>
            <Tab.Screen name="Stopwatch" component={Stopwatch}/>
            <Tab.Screen name="Timer" component={Timer} />
        </Tab.Navigator>
    );
}
```

Figure 11.29 – Our fresh AppTabs() functional component with Icon implemented

So, as we can see, the `screenOptions` prop works with a function that accepts the `navigation` and `route` props for each screen. Right now, we're using `route` as we want to check which screen is equal to each route. This way, we can have a case for every screen in our Bottom Tab Navigator. That function returns an object with a key called `tabBarIcon`, which has its value set to a function that's gathering a lot of information about the screen our user is currently focusing on.

This is where we check if the user is focused on a specific screen. Depending on that, we can render different types of icons. So, if the user is focusing on the Stopwatch screen, then we'll display a filled-in icon, while if they're not focusing, we'll only display the outline of that icon. This is a small detail that helps the user know when they're on the screen that they think they're on.

Now, setting the colors for our icons is a lot easier. For this, we'll use the `tabBarOptions` prop. We'll pass it an object containing two keys: `activeTintColor` for when the user is currently focused on that specific screen and `inactiveTintColor` for when the user is *not* currently focused on that specific screen.

Let's save this and check out our app! I think we can both agree that it looks 10x better now:

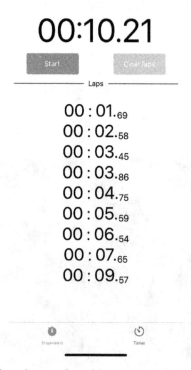

Figure 11.30 – Final app layout after adding icons to the Bottom Tab Navigator

Don't close the `routes.js` file yet! We still have one more thing to do. As we discussed in *Chapter 10*, *Building an Onboarding Screen*, the onboarding screen should only appear the first time you open the app. There is no reason to always see that onboarding screen. A lot of your users will be like, *OK, we get it, it's a Stopwatch app, just let me get to the Stopwatch part!*

How can we do that, though? This is where `AsyncStorage` comes in handy! For us to be able to use this package, we need to install it. Let's open our terminal again and write the following command:

```
expo install @react-native-async-storage/async-storage
```

Now that we've installed this cool little package, let's import it inside our `routes.js` file. Do that right after our `Icon` import, just like this:

```
import AsyncStorage from '@react-native-async-storage/async-storage';
```

Now, we can use this `Icon` import inside our `AppStack()` function. Once you have found that function, we should create a state variable called `viewedOnboarding`. This variable will help us know if the user has already seen the onboarding screen or not.

After defining that variable, we'd need to run a function right at the beginning of our mobile app. Do you remember the way we're supposed to that? That's right – another `useEffect()` function. I bet you're getting tired of these functions, but they're so great!

This `useEffect()` function should call another function called `checkOnboarding()`, whose purpose is to check if the user has seen the onboarding screen or not. Depending on that, we'll be setting our state variable, `viewedOnboarding`, to either `true` or `false`:

```
const [viewedOnboarding, setViewedOnboarding] = useState(false);

const checkOnboarding = async () => {
    try {
        const value = await AsyncStorage.getItem('@viewedOnboarding');

        if (value !== null) {
            setViewedOnboarding(true);
        }
    } catch (error) {
        console.log('Error @checkOnboarding: ', err);
    }
};

useEffect(() => {
    checkOnboarding()
}, []);
```

Figure 11.31 – Logic written for our AppStack() function

Now, we need another `async` function. But we can only use this package with an `async` function. We'll try to see if the local storage has that item stored and if it does, then we'll set the `viewedOnboarding` state variable to `true`.

You might be wondering when exactly we will add that item to our local storage. Well, we should do that when our user presses **Next** for the last time inside the onboarding screen. So, let's move on to the `Onboarding.js` file and make that happen.

Now that we're in the `Onboarding.js` file, we should begin by importing the `AsyncStorage` package again. After this, we should jump straight to the `scrollTo()` function. First, we'll make this function `async`. After that, we have an `if-else` statement. We'll change the *else* part, where we've had a `console.log()` living for no real reason and have the `navigation.navigate()` function instead. Let's see how we'll change that:

```
const scrollTo = async () => {
  if (currentIndex < slides.length - 1) {
    slidesRef.current.scrollToIndex({ index: currentIndex + 1 });
  } else {
    navigation.navigate('Tab Navigator');
    try {
      await AsyncStorage.setItem('@viewedOnboarding', 'true');
    } catch (error) {
      console.log("Error @setItem: ", err);
    }
  }
};
```

Figure 11.32 – Our modified scrollTo() function

Here, again, we'll try using a try-catch. As you can see, we're using `setItem` to set that item to `true` in the local storage. This is how this library knows that this item has been set to true in the storage.

Now, let's go back to the `routes.js` file. We're all set, but we need to make sure that we only display that route if our users have not seen the onboarding screen yet. We'll do that with **conditional rendering**, which is a technique we've been using since the beginning of this chapter. Let's see what that looks like:

```
return (
    <NavigationContainer>
        <Stack.Navigator>
            {!viewedOnboarding && (<Stack.Screen name="Onboarding" component={Onboarding} options={{
                    headerShown: false
                }} />)}
            <Stack.Screen name="Tab Navigator" component={AppTabs} options={{
                headerShown: false
            }} />
        </Stack.Navigator>
    </NavigationContainer>
);
```

Figure 11.33 – Conditional rendering applied to our <Stack.Screen /> onboarding component

As you can see, we're checking for our state variable, `viewedOnboarding`. If this has been set to `false`, that means our user has not seen the onboarding screen yet, so our route should be displayed. If this is set to `true`, that means we're not going to display any route, practically making the onboarding non-existent.

And with that, we're done with this app! Save all your files, reload your JavaScript, and take a look at your app. At first, you'll see the onboarding screen. Tap next until it disappears and play around with the Stopwatch and Timer screens. After that, open the app again and you'll see something amazing – the onboarding screen is not displayed anymore! Instead, you'll be transported straight to the Stopwatch screen – more exactly, the Tab Navigator screen.

Congratulations! You now have a pretty cool and functional app at your disposal. Go ahead and brag about it to your friends and family; let them see how much you've progressed!

Summary

This chapter has been a long journey for both of us. Fear not! The greater the challenge, the better the reward. You have reached the end of a pretty long and interesting journey. You now have a fully functional app that you can show to your friends. These are the first stepping stones of your journey toward becoming a great React Native developer.

We've started this chapter by looking at React Navigation. Creating routes and linking them to the onboarding screen was one of the coolest things we've ever done. This was also incredibly easy, which proves once again how great the React Native community is.

After linking our app to the React Navigation library, we started working on the Stopwatch screen. We learned that the `setInterval()` function is not that accurate, due to which we started working with date objects, which proved to be a lot more efficient for keeping time.

Finishing the Stopwatch screen felt like a big win, so because of that, creating our Timer screen went a lot smoother. But yet again, we learned something new and that was how to play a sound after the timer finished running. I hope doing this opened a lot of doors of creativity for you.

At the end of this chapter, we focused on the user experience and made sure that the user will see some icons whenever they're looking at the Bottom Tab Navigator. On top of that, we worked with a library called `AsyncStorage` so that we could keep the onboarding screen away from an already experienced user.

Learning so many things was a breath of fresh air. Yes, this was a lot of information, but I hope you realize how important it is to tackle as many challenges as possible. Just like in real life, they help us build experience, which helps us become great programmers.

Now, let's get ready for the next chapter, where we're going to discuss what other paths you can take as a React Native developer to become a great programmer.

12
Where To Go from Here?

Over the last 11 chapters, we have been learning a lot about React Native and how to build cross-platform mobile applications with it. We also saw how much time we can save with libraries such as Galio, which can quickly help us build cool-looking apps with just the default components.

In our React Native journey, we tackled many different situations where we learned about the user interface, user experience, Expo, and a lot more. I hope you've enjoyed everything we've learned and that you're excited to learn about the many ways you can continue learning about and studying React Native.

This chapter is going to focus on how we can start accumulating more knowledge. We'll discuss how we should attack any new challenges we might face in the future whenever we want to learn something new or just use small libraries for our app development.

We'll also discuss Galio's community and what you can do to join and help develop Galio. We'll also learn about the importance of open source for the programmer community and how it can help you with your career.

I'll also leave you with some tips and tricks you should use whenever life isn't easy while developing your cross-platform mobile applications. I hope you will revisit this section often to gain some inspiration and motivation for your work.

Last but not least, we'll discuss the importance of books and reading. More specifically, we'll discuss why it is so important for us to try and learn as many things as possible from a multitude of sources.

I hope you enjoy this chapter and that it gives you the morale boost you need to continue learning and having fun with React Native.

This chapter will cover the following topics:

- Always read the documentation
- Galio's community
- Tips and tricks

Always read the documentation

Every time you start a new project or tackle a new programming challenge, it is important to make sure you're prepared before actually diving in and starting to code. This planning and preparation can come in many forms, but one of the most important things you can do is read the documentation of the technologies you're working with.

Now, the thing with reading the documentation is learning how to comprehend it. In time, with experience, you'll understand what makes great documentation and what exactly you wish to gain from it.

Reading technical documentation might not always be easy, especially for a beginner programmer. Some terms that are used in one language might be different compared to another language. Documentation also contains a lot of jargon terms that you might not feel comfortable with at the beginning of your journey.

Fear not! I've built a list of tips for reading and solving the challenges you'll face in the future:

- **Chill out!**: Reading the documentation takes time; it's just like reading any book. Don't hurry straight to the middle of the book to read about all the fighting scenes; sometimes, it's better to read the introduction so that you can familiarize yourself with every character's ability. So, get ready for a cool little ride and enjoy every part of it. If you ever get tired, just take a break, look out the window, and come back later with a clearer head.

- **Read multiple sources, like a real journalist**: There will be times when the documentation you're reading might be a bit too advanced for you or simply incomplete. Sometimes, you might even encounter some passages you can't understand. Therefore, you'll need to learn how to move even further than the official documentation. Read more articles about that subject from different sources. This will help you refresh yourself on the concepts you're learning about by seeing different perspectives and examples.

- **Review terms**: You'll always see some new term you've never heard of before. You should make a list of all the terms you don't understand and take some time to review them. This will help you in the long run, especially while you're trying to learn about more libraries.

- **Version check**: There are always multiple versions of the library you're trying to study, so make sure you're reading the correct documentation!

The reason why you should always go to the technical documentation instead of video tutorials first is because, usually, the latter has a very superficial level of explaining concepts. Of course, there are great video tutorials out there, but it's always better to learn straight from the source and, as I mentioned previously, then check out other sources.

There are sometimes two parts of any documentation you can find online: *Getting Started* or *Guides* and *Docs* or *Documentation*. The first one will always have something like a small, simple example of how to use the package or library. The purpose of this part is for you to understand as much as possible about the context of the package. The *Documentation* part is more like a phone book. It is straightforward and you can always find specific information about a specific thing, which means that this part is not a guide on how to make sure you're installing or using it correctly – it is more a dictionary filled with definitions for every keyword you'd need.

Now that we've talked about reading the documentation, we should discuss reading books. A book can be kind of like the *Getting started* part of a big documentation project but with more examples and practical challenges.

Books are really helpful

Books have the purpose of guiding you through your journey of learning a specific skill, or stories of different superheroes fighting bad hackers for a better world. But we all know we're talking about technical books here, so let's ignore the superheroes for a second.

By reading a book, you're not replacing the documentation as the information might be outdated sometimes or too opinionated. However, as I've already mentioned, this doesn't mean that a book wouldn't be helpful to you.

It's important to read as many books or articles as possible, especially on the subject you're the most interested in. Getting as many opinions and workflows as possible will make you a better programmer.

Why do I say a better programmer? I believe that the most important aspect of a programmer's life is the level of experience they've garnered by researching, practicing, and experimenting with different technologies or libraries.

Books can help you get there. There are a lot of famous programmers with strong opinions regarding the way you should write code or even think about it. They've all put their knowledge into books and because of that, we should make sure to use their knowledge to our advantage.

This is exactly how progress is made in any scientific domain. By absorbing what others have thought and building upon it, you'll have the chance to create something new.

This is why books are really helpful for us programmers – they are snapshots of knowledge. We use them to stay in touch with technologies and ways of thinking.

But that doesn't mean that we can gain all the knowledge necessary for our progress just by reading books. They are more like extra guidance. The important thing is to try and gain information from as many sources as possible, and books are one of those sources that could inspire you more than other sources besides the original documentation.

Galio's community

If you've ever felt like there's no place for you… well, you should try Galio's Discord. It'll help you with whatever specific questions you might have.

We're not here to discuss just Discord – we're here to discuss how Galio's community might help you learn the basics of helping out with open source projects and how this whole experience of helping Galio out can be beneficial to you belonging to a community.

There are multiple ways of getting involved with Galio. In this section, we'll discuss all the ways you can learn about Galio, engage with the community, and also study the way it works. Let's start by talking about Galio's website.

Galio's website

We've started with Galio's website because this is usually the first or second thing people find whenever they're looking for Galio. You can find their website by navigating to `https://galio.io/`.

First of all, the website looks pretty cool:

Figure 12.1 – Galio's website

As you can see, Galio's color palette is present on the website as well. This is where everything starts and where there are links to everything regarding Galio. This is where people first begin understanding how Galio works and why it has been created.

In the navbar, you can find different links to help you with your journey. The second one is for **Starter Kit**. This will take you straight to a GitHub repository where you can try out a project full of different screens built with Galio. This will help inspire you regarding what Galio can do to help you with your project. There are lots of example screens you can learn from or even use in your projects as everything is open source and free to use, reuse, and modify as you please.

The third link, **Component**, takes you straight to the documentation, something that we'll tackle later in this chapter. Then comes the fourth and fifth links: **Examples** and **Premium Themes**. The first one has the purpose of showcasing what others have built with Galio's library; it reaches out to the community and helps them by showcasing their work. The second one is there to help other developers buy themes built with Galio to maximize their workflow and boost their productivity. This is for the developers that have already worked with Galio and want to build something really quick and of premium quality.

You should visit the website and take a look around to see what else you can learn and discover about Galio. Now, let's move on to Galio's documentation.

Galio's documentation

This is where everybody comes to study and learn about how to use Galio, as well as understand every little aspect of Galio. Even though Galio has a guide, this documentation is more like straight technical documentation:

Figure 12.2 – Galio's documentation website

When you head to `https://galio.io/docs/`, you'll notice a big landing page containing some information about Galio. You'll have to scroll down to see the navbar and start embracing Galio's documentation.

You should pay attention to the navbar as this is where you can find all of Galio's components, as well as lots of information about each one.

Let's check out the documentation for the `<Block />` component:

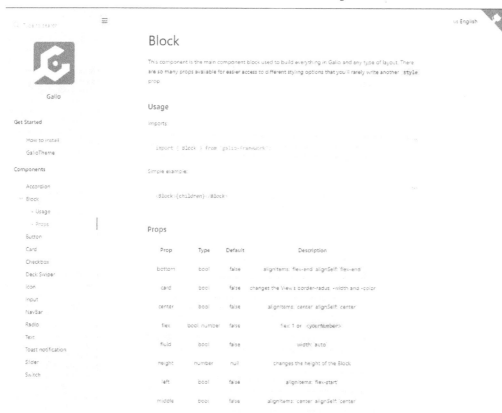

Figure 12.3 – Block component overview

As you can see, every component has a description for you to understand more about it. Then, you have an example of how to use it and import it into your project. There's also a table full of props that you can use with the component that you're interested in.

Here, you can information about those props. There's also a description telling you what type of styles it applies or what it does.

You can find a lot of information about Galio's component by diving straight into the code in Galio's GitHub repository.

Galio's repository

You can find Galio's GitHub repository at `https://github.com/galio-org/galio`. You can find lots of things here. Here, the most powerful thing you can do is check out Galio's sources code:

Figure 12.4 – Galio's GitHub repository

By doing that, you'll learn how Galio has been created and also debug your code in case something is not working the way you were expecting. This is the perfect place to take inspiration for a UI library or open source project.

There is also a **Wiki** tab. This is where you can find a lot of extra find information about Galio. By extra information, I mean stuff such as the status of the development, how to use it, and also a guide on how to contribute to this open source project.

There are lots of ways to contribute to Galio and I support you discovering your own path to helping out with the library.

The community is what truly supports Galio. Without the community, we wouldn't be able to progress as much as we have, and we're always there to embrace you if you need help.

This is what it means to be part of a community. It means to help and be helped. It means to believe in a project so much that you're willing to support it as much as possible. And if you do feel like Galio deserves your support, jump on board our ship and let's work together!

Galio's Discord

A good place to start interacting with people, besides the **Issues** tab when asking for help or solving bugs, is Discord. You can find the Discord link on our website.

On Discord, everybody is sharing funny pictures or asking questions regarding how to use Galio. It's like having a small online family, always helping you out with your Galio issues.

Now that we've been through all of this, let's look at some tips and tricks for your React Native projects.

Tips and tricks

React Native is great, but all great things have some minor flaws. Because you'll never know what type of error you might encounter, I've decided to create a list of the most common errors and fixes. Let's begin!

Import error

This error usually comes up whenever you've mixed default and named imports. Let's check out the error message:

```
Invariant Violation: Element type is invalid: expected a string
(for built-in components) or a class/function (for composite
components) but got: undefined. You likely forgot to export
your component from the file it's defined in, or you might have
```

```
mixed up default and named imports.
Check the render method of 'App'.
```

As we can see, this error has been caused by a component being imported into the main App.js file. Unfortunately, the error message does not tell you which component or line is breaking the app. To make sure this won't happen again, you'll have to double-check your exported/imported components to ensure there are no mistakes. Now, let's see how this could have happened.

We know that there are default and named imports, but let's discuss the differences between them.

Let's assume you have the following export:

```
export const myComponent;
```

Now, this is a named export. Because it's a named export, you'd have to import it, like so:

```
import { myComponent } from './myComponent';
```

Now, let's see how a default export works:

```
export default myComponent;
```

Being a default export, you're allowed to import it without the curly braces. Another cool thing is that the name doesn't matter anymore:

```
import stillMyComponent from './file';
import myComponent from './file';
```

These two imports work the same, even though we have named them differently. This is pretty cool, right? The more you know, the more prepared you are.

React Native version mismatch

So, let's get straight into it and check out the error message:

```
React Native version mismatch.
Javascript version: X.XX.X
Native version: X.XX.X
Make sure you have rebuilt the native code...
```

This shows up whenever you're trying to build the app. This is because the bundler you're using when you're using the `expo start` command inside your terminal is using a different JavaScript version of `react-native`. You may encounter this after you've upgraded your React Native or Expo SDK version, or even when you're trying to connect to the wrong local development server.

Let's fix this. Begin by closing the Expo server. After that, you should try two different things: the first is to remove the `sdkVersion` file from your `app.json` file. The second is to make sure that the file matches the value of the `expo` dependency in your `package.json` file.

By having a managed workflow project, you can make sure your `react-native` version is correct by running the `expo upgrade` command. If you have a bare workflow project, make sure you've upgraded everything correctly.

Once you've done everything, you should clear your caches by running the following command:

```
rm -rf node_modules && npm cache clean --force && npm install
&& watchman watch-del-all && rm -rf $TMPDIR/haste-map-* && rm
-rf $TMPDIR/metro-cache && expo start --clear
```

Now, we shouldn't see the error again – that's great!

Unable to resolve

The error message for this is going to be something like the following:

```
Unable to resolve module <module name> from <path>: Module does
not exist in the module map or in these directories
```

This error is usually generated by using symbols such as ^ or ~ in your `package.json` file.

To solve this, remove those symbols and delete your `node_modules` folder. Upon reinstalling all the packages, everything should work fine.

Summary

In this chapter, we got ready for our lives as React Native developers. We discussed a lot of things that should help you in your journey. I also strongly believe I've been able to help you out by inspiring you to pursue knowledge as much as possible.

First, we discussed how helpful the documentation is. We also learned how to gather information from as many resources as possible. Books are a really important part of our education, so make sure to at least try reading some more books on React Native.

Then, we discussed Galio and how we can get in touch with the community. We saw how many resources we have at our disposal, free to use and also of great quality. This will be helpful when we meet again (at least I hope so) on Galio's Discord or repository.

After this, we tackled some common React Native issues and learned how to fix them. I hope you found this helpful and that you'll come back to these later so that you can fix errors quicker than performing a Google search.

At this point, you're ready to start developing projects. You're finally ready to come up with an idea and work toward making it a great success. I hope this book has helped you and that you've learned as much as possible. I also hope that you're a lot more hopeful and hyped for the future. Stay safe and healthy!

Packt.com

Subscribe to our online digital library for full access to over 7,000 books and videos, as well as industry leading tools to help you plan your personal development and advance your career. For more information, please visit our website.

Why subscribe?

- Spend less time learning and more time coding with practical eBooks and Videos from over 4,000 industry professionals

- Improve your learning with Skill Plans built especially for you

- Get a free eBook or video every month

- Fully searchable for easy access to vital information

- Copy and paste, print, and bookmark content

Did you know that Packt offers eBook versions of every book published, with PDF and ePub files available? You can upgrade to the eBook version at packt.com and as a print book customer, you are entitled to a discount on the eBook copy. Get in touch with us at customercare@packtpub.com for more details.

At www.packt.com, you can also read a collection of free technical articles, sign up for a range of free newsletters, and receive exclusive discounts and offers on Packt books and eBooks.

Other Books You May Enjoy

If you enjoyed this book, you may be interested in these other books by Packt:

React and React Native - Third Edition

Adam Boduch , Roy Derks

ISBN: 978-1-83921-114-0

- Delve into the React architecture, component properties, state, and context
- Get to grips with React Hooks for handling functions and components
- Implement code splitting in React using lazy components and Suspense
- Build robust user interfaces for mobile and desktop apps using Material-UI
- Write shared components for Android and iOS mobile apps using React Native

React Native Cookbook

Daniel Ward

ISBN: 978-1-78899-192-6

- Build UI features and components using React Native
- Create advanced animations for UI components
- Develop universal apps that run on phones and tablets
- Leverage Redux to manage application flow and data
- Expose both custom native UI components and application logic to React Native
- Employ open source third-party plugins to create React Native apps

Packt is searching for authors like you

If you're interested in becoming an author for Packt, please visit `authors.packtpub.com` and apply today. We have worked with thousands of developers and tech professionals, just like you, to help them share their insight with the global tech community. You can make a general application, apply for a specific hot topic that we are recruiting an author for, or submit your own idea.

Share Your Thoughts

Hi!

I am Alin Gheorghe, author of *Lightning-Fast Mobile App Development with Galio*. I really hope you enjoyed reading this book and found it useful for increasing your productivity and efficiency in Galio.

It would really help me (and other potential readers!) if you could leave a review on Amazon sharing your thoughts on *Lightning-Fast Mobile App Development with Galio*.

Go to the link below or scan the QR code to leave your review:

https://packt.link/r/1801073163

Your review will help me to understand what's worked well in this book, and what could be improved upon for future editions, so it really is appreciated.

Best Wishes,

Alin Gheorghe

Index